The Essentials of AutoCAD

Matthew M. Whiteacre

Texas A&M University

KENDALL/HUNT PUBLISHING COMPANY
4050 Westmark Drive Dubuque, Iowa 52002

ISBN 0-7872-9385-7

Printed in the United States of America

10 9 8 7 6 5 4 3 2 1

Contents

Introduction

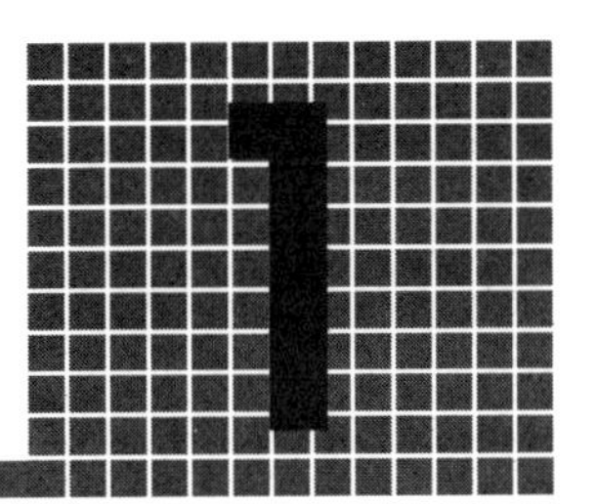

Background

Over the last two decades, computer graphics programs have come a long way. What used to be the realm of large super-computers, is now commonplace on home PC's. When the movie "The Last Starfighter" was produced, the animation was done strictly by computer; no models, no hand tweening, strictly on computer. This was the first movie to be so produced. The computer used was a Cray YMP and the cost was phenomenal. Today, Saturday morning cartoons can be produced on PC for a tiny fraction of the cost. There are many new programs that have been written during these two decades to meet the changing demands of industry and the public. One of the programs that has been around for the entire two decade period is AutoCAD.

AutoCAD was released in 1982 and was one of the first computer graphics programs to be written for a PC and marketed to a commercial environment. The initial release was crude by today's standards, but in 1982 the program was impressive. It was designed to allow professional draftsmen to create and edit drawings rapidly. The drawings were of the same nature the draftsmen had been working with, just the method to create them was different. Instead of picking up a pencil and paper, they executed the program on their PC and worked in that media. The concept of using a computer to do basic drawings was not new, but the concept of using a personal desktop computer to do it was very different.

AutoCAD did not try to design a program that was especially well suited for engineers, nor one that was especially tailored to architects: their goal was to make a simple to use generic program. They allowed third parties to develop software to compliment and enhance their basic program, thus you could get a separate module to draw structural steel, of windows and sofits, or landscaping contours, or piping design. You could even write your own enhancements for the software using a programming language called LISP. Using these basic concepts, AutoCAD gained a dominant place in the market and has held that place for 20 years. Today, of the people who use a personal computer drawing program to produce commercial drawings, 70% of them use AutoCAD.

Purpose of This Book

The intention of this book is not to create expert users of AutoCAD. It is, instead, to provide an introduction to the basics of AutoCAD in a simple easy to follow manner. The sequence of instruction will have you up and using AutoCAD quickly, and will then build on the foundations as the book progresses. By the end of Chapter 2, you should feel comfortable getting into AutoCAD and doing a basic drawing. Some concepts, which are important in the long run, will be lightly brushed over initially to get you up and running. These concepts will then be re-examined in subsequent chapters to round out your abilities in AutoCAD.

What Is CAD?

Before beginning the study of a new program, it is important to realize what the purpose of the program is and how it can be applied. If you want to write a report on a subject, you want to use a word processor, not a spreadsheet. Likewise, if you want to create a drawing you use a drawing program, not a word processor. The problem is that there are many types of drawing programs with different focuses. Some focus on creating photo-realistic images, some on editing images, some on illustrations, and some on technical drawings. Each has a place in industry.

AutoCAD falls into the technical drawing category (actually if you use the 3D capabilities of AutoCAD, it can produce photo-realistic images). It will not apply bright, bold splashes of color on a drawing, instead it concentrates on allowing you to draw geometric figures (lines, circles, arcs, etc.) of precise size and then document, with dimensions, what you have drawn. Using this, you can lay out orthographic views of objects, draw pictorials, or even solve problems using descriptive geometry.

Conventions Used in this Book

While discussing actual AutoCAD commands in the text they will be listed in all caps, (i.e. LINE). Command streams will be shown in Courier font. The user input will be in boldface while directions will be shown italicized and enclosed in braces.

```
Command: offset
Specify offset distance or [Through] <1.0000>: 1/12
Select object to offset or <exit>: {select the circle}
Specify point on side to offset: {select a point inside the
      circle}
Select object to offset or <exit>:
```

Dedication

I would like to publicly thank several groups of people who have greatly assisted me in writing this book. First of all has to be my family, Jari, Monica, Sylvia, Alicia, and John Paul, who put up with my being at the office many evenings and weekends to complete this book. Secondly is a group of friends, Mike, Jeff, and Jim, who help me make my deadlines by constant harassment about my progress. Without the support of either of these groups progress on the book would have been much slower and more painful than it was.

Finally I would like to thank Dr. Gerald Vinson for convincing me that I could write a book and encouraging me to continue writing.

Getting A CLUE

Objectives

After completing this chapter, you should be able to:

1. Enter AutoCAD, Exit AutoCAD, Save a drawing file
2. Create drawings using Lines, Arcs, and Circles
3. Erase items from a drawing
4. Use the Undo command
5. Select items using either a normal window or a crossing window
6. Add text to a drawing using MTEXT or DTEXT

Beginning a New Drawing

After initially launching AutoCAD 2002, you will be presented with a startup dialog box shown in Figure 2.1. (If your box does not look like this, see Appendix 5 for possible solutions.)

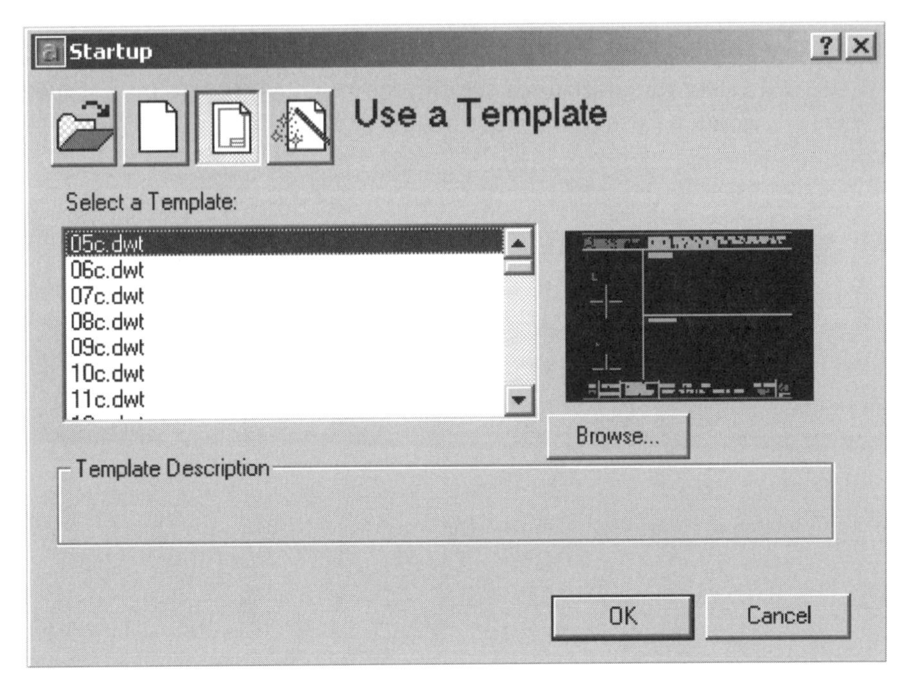

Figure 2.1

The various options are:

Open an existing drawing file

Start a new file from scratch

Use a template file

Use AutoCAD's drawing setup wizard

Starting a file from scratch will load a basic drawing using the default setting established by AutoCAD. If you want to load a different set of parameters, then you must use a template file. This is a basic drawing file with the parameters set to whatever style you want. The creation of template files will be discussed in chapter 12, for now you will either be using the standard new file or a template file provided by your instructor.

AutoCAD's setup wizard will ask you basic questions about the drawing you would like to create, and then attempts to create a drawing using the parameters you specified. When just beginning, this is a tempting option, however since you do not really know what settings you will need, it has its drawbacks. Once you understand all that needs to be done to create a template, then there is little need for the wizard.

Now that we have seen a little about beginning a drawing, let's load AutoCAD and start a new drawing from scratch.

1. Double click on the AutoCAD icon to launch the program.
2. Select the start from scratch icon or use the create a drawing tab and select from scratch. You should get a screen which looks like Figure 2.2.

Figure 2.2

The majority of the screen is reserved for the actual creation of the drawing. The very top has pull down menus common to most Windows programs. The very bottom has a status line which shows what modes are active at the current time and a location of the cursor in X, Y, and Z. Just above this is the command prompt area where you can enter commands directly to AutoCAD. Set on the left edge and just above the drawing area are toolbars to facilitate drawing creation and editing.

Now that we have a drawing open, you can begin to add objects (the term AutoCAD uses for lines, arcs, etc.) to create a drawing. The normal method for adding objects to a drawing is by using the mouse, so we should examine the mouse before going any further.

Using the Mouse

Mice come in many different sizes and flavors, but the one reasonably common thing about them is that they all have at least 2 buttons, one labeled left and one labeled right. When using AutoCAD, the right button will be referred to as the right button and the left button is referred to as the pick button. The right button has the function of activating context sensitive menus or mimicking the function of the enter key, depending on what state AutoCAD is in at the time. The left button is used to pick icons, menu items, or points on the drawing area.

By placing the mouse over a toolbar icon and pressing pick (i.e. the left button), AutoCAD will activate the command associated with that icon; pressing the right button will activate a selection menu of all toolbars available in AutoCAD. The pulldown menus are activated using the left button. Point selection within the drawing area is made with the pick button, however, left clicking within the drawing area may also be used to select objects for editing. AutoCAD is context responsive to the needs of the user as it perceives them.

Actually Creating a Drawing

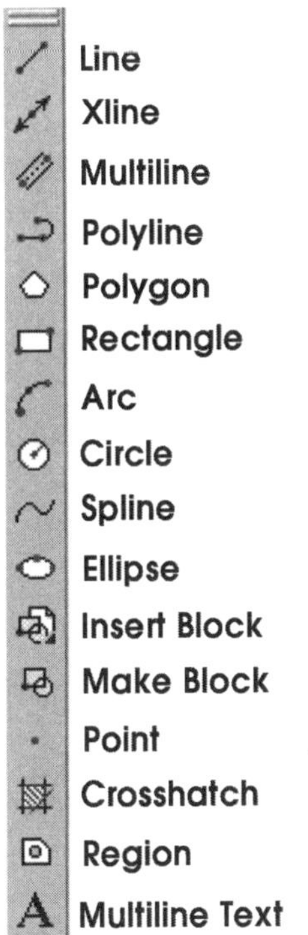

Figure 2.3

The toolbar on the far left edge of the screen is the Drawing toolbar. Figure 2.3 shows this with the buttons labeled. For this chapter, we are going to focus on the basic three commands: Line, Circle, and Arc. The basic purpose of each of these commands should be readily identifiable, but a brief look at how AutoCAD implements them will be useful.

Drawing Lines

The line command allows you to draw independent straight line segments from point to point in AutoCAD. Once a line segment is created it becomes an object and can be manipulated independently from any other object on the screen. The command used to execute the line command is LINE if you are typing commands via the keyboard, or the alias L can be used also. Each line segment end point will be selected using the left mouse button (the pick button). A session would look like:

```
Command: line
Specify first point:
Specify next point or [Undo]:
Specify next point or [Undo]:
Specify next point or [Close/Undo]:
Specify next point or [Close/Undo]:
Specify next point or [Close/Undo]:
Specify next point or [Close/Undo]:
```

Notice that for the fourth and subsequent points, AutoCAD offered the option "Close". Using that option will draw another line segment from the previous point to the point selected as the "First point:" for the current set of line segments. To use that option you must type "Close" or "C" through the keyboard and press enter.

The other option available is "Undo" or "U" which will unselect the previous point and allow you to place it in the correct location. There are no limits on the number of times you may use the undo option. You could undo all the selections back to the "First point:" prompt if you wanted to.

To stop drawing lines with the current command you must press Enter or Escape to terminate the current command. When you do this, AutoCAD will return to the "Command:" prompt awaiting further instructions.

To access all those options with the mouse, right click during the line command and AutoCAD will display the popup menu shown in Figure 2.4. From here you may select any of the options listed above, or execute a Pan or Zoom operation. These will be discussed in chapter 4.

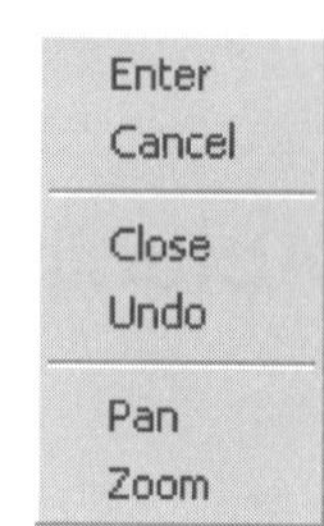

Figure 2.4

Drawing Circles

The circle command allows you to draw circles in a variety of ways. The most common form requests a center point and the radius of the circle. With this information, AutoCAD draws the circle and terminates the command, returning to the "Command:" prompt. To use the command through the keyboard type "Circle" or "C" and press enter.

```
Command: c CIRCLE
Specify center point for circle or [3P/2P/Ttr (tan tan radius)]:
Specify radius of circle or [Diameter]:

Command: CIRCLE
Specify center point for circle or [3P/2P/Ttr (tan tan radius)]:
Specify radius of circle or [Diameter] <0.9059>: D
Specify diameter of circle <1.8117>: 3.25
```

In the above set of commands, the first set simply used the defaults and draws a circle using the center point and radius as selected by the left mouse button. The second circle selects the center point with the mouse, but utilizes the keyboard to specify that the diameter should be used and then that diameter is typed via the keyboard.

The other options that exist for circles are:

3Point—allows you to select any three non-colinear points and draws a circle through those points

2Point—allows you to specify two points along a diameter of a circle and draws the circle

TTR (tan tan radius)—does not ask for points to locate the circle, but two objects already drawn on the screen. A circle will be drawn tangent to those 2 objects with a specified radius (if the resulting circle does not exist, the AutoCAD will ignore it and will go back to the command prompt).

As with the Line command, you may right click and get a popup menu to select the various options from. The two possible menus are shown in Figure 2.5. The left most is displayed if you right click before selecting a center point and the right most if you have selected a center point.

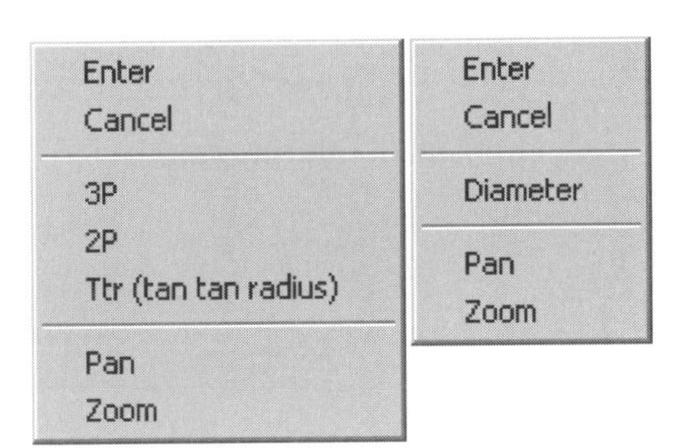

Figure 2.5

Drawing of Arcs

Drawing arcs in AutoCAD requires a little more planning than either lines or circles. There are many more options for creating arcs than for lines and circles combined. Given this, we will explore another method to select commands in AutoCAD, the pulldown menus.

Under the Draw pulldown, you can select lines or circles, however these are easily drawn using the icons. Arcs present more of a challenge, since in normal engineering drawing, the default selection via the icon (a 3 point arc) is not the most common. The pulldown menu for drawing arcs is shown in Figure 2.6. The options available, in various combinations are:

Start: The point at which the arc should start.

Center: The center point of the arc.

End: The end point of the arc.

Angle: The total angle included in the arc.

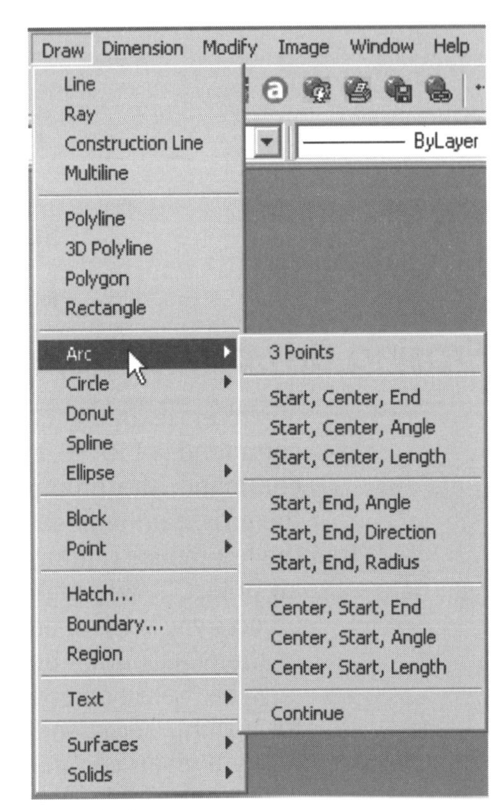

Figure 2.6

Length: The length of the chord defined by the arc. This is not arc length, but chordal length.

Direction: The tangent direction for the arc at the start point.

Radius: The radius of the arc.

Arcs are always drawn counterclockwise in AutoCAD. If you want to draw an arc clockwise, you must use the Angle option and specify a negative angle. Figure 2.7 shows various options of the arc command.

Using these three commands there are many things you can draw, but this is just a glimpse of the full power of AutoCAD. The ability to draw/create items in a CAD system is secondary to the ability to edit and make corrections. That is where the true power of a computer aided design system lies.

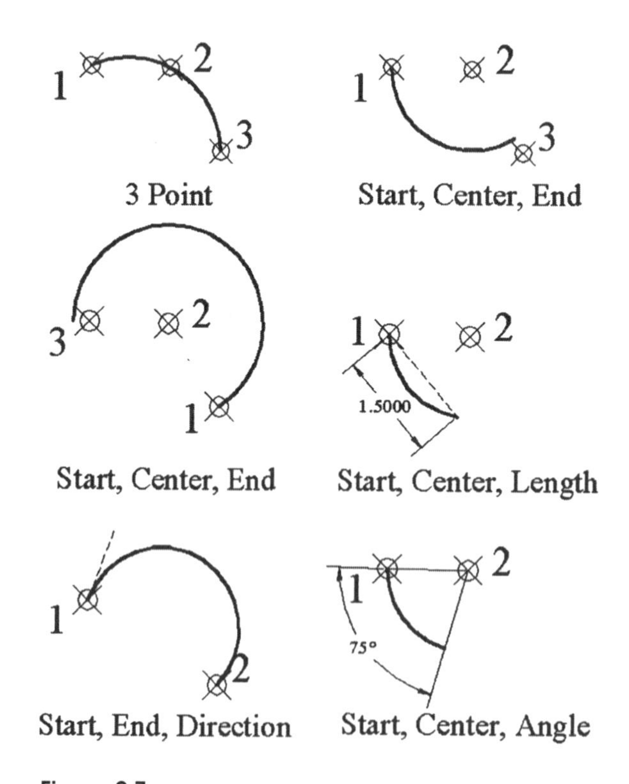

Figure 2.7

Erasing Unwanted Objects

To erase an object from the drawing you must invoke the ERASE command. The command is ERASE, or E is sufficient or pick the ERASE icon. Once you are using the command, the prompt in the command area will change to "Select objects:" and the crosshairs will change to a small square, indicating that you should pick (using the left mouse button) the objects you wish to delete from the drawing. As you select an object, it will highlight (become fuzzy looking). Once you have selected all the objects you wish to erase, complete the command by pressing the right mouse button or the enter button on the keyboard.

The "Select objects" prompt is very common in AutoCAD and offers many methods for forming *Selection Sets*. A Selection set is a collection of objects that are to be acted on by a given command. Initially, three options of the selection objects prompt will be examined: object picking, Window, and Crossing.

Object Picking requires that you place the small square cursor over an object and press the left mouse button. It will select one object and only one object. In order to select more than one item, you should use the Window or Crossing option. To access these options, place the square cursor on a portion of the screen without an object and press the left mouse button. AutoCAD will respond by telling you to "Specify opposite corner:" of a rectangular region. This region will either be a Window area or a Crossing area depending on your next action. By moving the mouse to the right or left of the initial selection point (vertical motion does not matter) you can switch from a Window to a Crossing. Windows are to the right and Crossings are to the left. You will notice that the rectangular area will switch from a solid rectangle to a dotted rectangle to illustrate the shift from Window to Crossing.

If you choose a Window area, then all objects which are completely enclosed by the rectangular region will be selected. While a Crossing area will select all those objects completely within the rectangle, it also selects those which touch the area at any point. In Figure 2.8 below, if the area being selected was a window, then the line and arc would be selected, but not the circle; if it were a crossing, then all three objects would be selected.

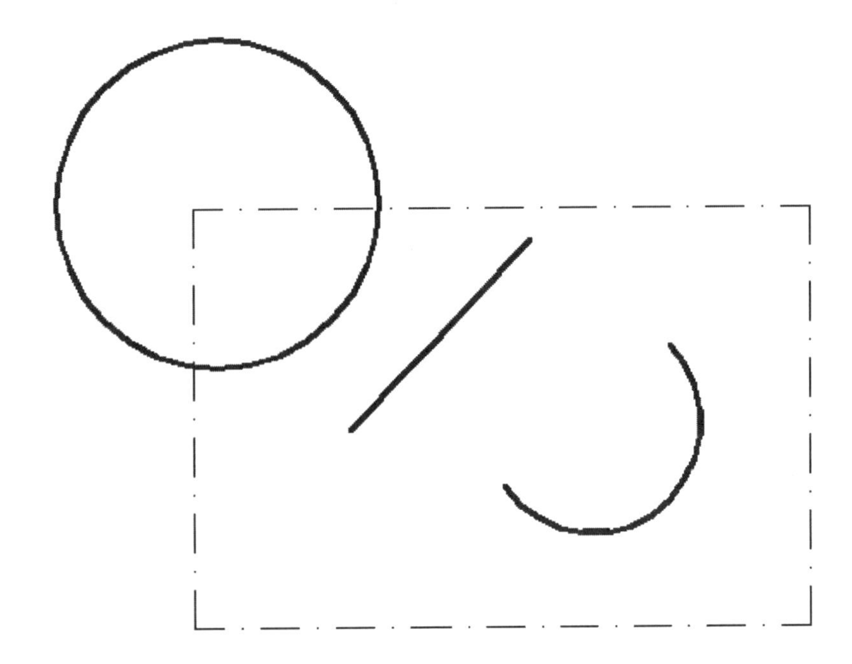

Figure 2.8

To remove an object from the selection before completing the erase, hold the shift key down while making a selection. This will unhighlight those objects.

Undoing an Action

AutoCAD, like most Windows programs, has a command specifically designed to give you an escape route when you have just done something you really didn't want to do. The Undo command. The icon is located on the upper toolbar or through the keyboard the command is "U" and it will reverse the last action taken. By using "U" multiple times you can, in theory, undo your entire drawing session. The UNDO list begins when you open a drawing, so any modifications you made during the previous drawing session cannot be undone.

A Bit About the Title of the Chapter

This chapter is titled "Getting A CLUE". The phrase "ACLUE" is an easy acronym for remembering the keyboard shortcuts to the five commands described in the chapter. **A**rc **C**ircle **L**ine **U**ndo **E**rase. You will find that it is frequently easier to use the keyboard shortcuts with your non-mouse hand than it is to run the mouse over to the appropriate toolbar and select an icon or to use the pulldown menus.

Basic Settings in AutoCAD

Objectives

1. Understand the importance of grid and snap
2. Understand the function of ortho and polar
3. Be able to control the size of the crosshairs on screen
4. Be able to use multiple layers in a drawing
5. Be able to use Template files
6. Be able to use Object Snap
7. Use the Standard layer Structure
8. Be able to activate new toolbars

Grid and Snap

Grid and snap are actually very different items in AutoCAD. Both can be controlled through the status line at the bottom of the AutoCAD screen (shown in Figure 3.1). In the example shown, Snap is activated (referred to as turned on) and Grid is turned off. The state of either can be controlled by clicking on their button.

-1.5000, -7.0000 , 0.0000 | SNAP GRID ORTHO POLAR OSNAP OTRACK LWT MODEL

Figure 3.1

In the final analysis, grid is not a critical concept and many people do not even use grid while drawing. Grid is simply the background set of dots that AutoCAD presents to assist you in measuring distances. The grid will not print. If you want a background grid on your prints, then special measures must be taken to draw that grid in AutoCAD.

Snap, on the other hand, is one of the most critical concepts for a beginning user of AutoCAD. Snap is a translation device that allow a computer and a human to speak on equal terms about precision drawing. The computer would prefer to think all input was accurate to 8 or more significant digits, while a human, looking at a standard monitor, would have trouble getting more than 3 significant digits (there are about 1000 pixels across a standard screen). By judicious setting of snap, this problem can be overcome. Snap requires that the cursor lock in at specific intervals.

Initially, this gives an uncomfortable feeling since the crosshairs on AutoCAD seem to jump from point to point (they really do jump). You are left with the question, what happens if I need to select a point which is not on a snap interval? You can either type the exact X and Y value through the keyboard or change snap to a more convenient interval. A common error beginning students make is to turn snap off to select a point and then forget to turn it back on. A good operating principle is DO NOT **DRAW** ANYTHING WITH SNAP TURNED OFF. When editing a drawing (using Erase, for instance) turning snap off is perfectly acceptable, just remember to turn it back on before drawing.

To control the setting of either Snap or Grid, you should right click on either button and select the Settings option from the popup menu displayed. This will activate the Drafting Settings dialog box shown in Figure 3.2.

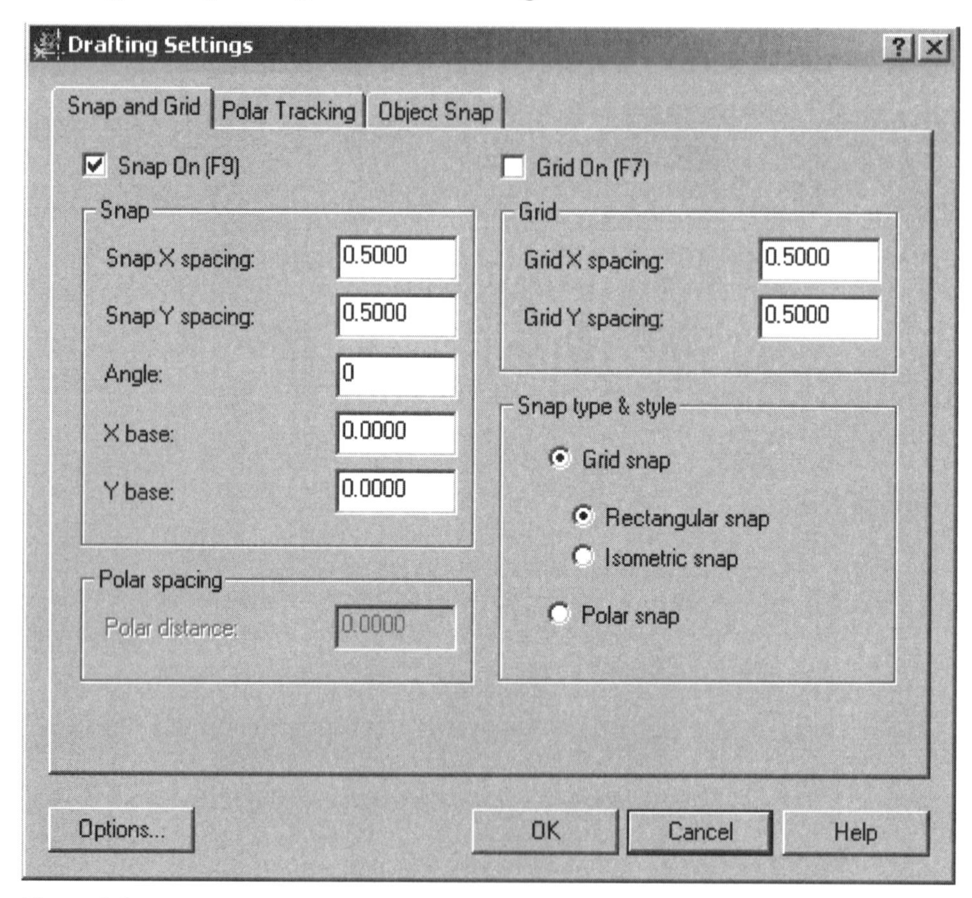

Figure 3.2

You may change either setting by modifying the value shown in the dialog box. Alternatively, you can type the command GRID or SNAP in at the Command: prompt and modify the value through the keyboard.

Use of Ortho

Right next to the Grid and Snap buttons on the status line is a button labeled ortho. This is a special mode AutoCAD uses to restrict the location of point selection when drawing objects. If activated, after the initial point is selected, subsequent points are restricted to being directly above, below, to the right or left of the previous point. This allows you to draw stair steps with line segments, but not ski slopes. It can be turned on or off at any time by left clicking on the button labeled ORTHO.

Use of Polar

Similar to Ortho, polar allows you to draw lines at predetermined angles. It is mutually exclusive with Ortho, so only one of them can be active at any time. By default the angles for Polar are set at 90 degrees. Even at this setting Polar has advantages over Ortho. With Polar, the crosshairs will not be forced to constrain themselves to the fixed angles, but will lock in to the angles when the crosshairs get within 3 degrees of the angle settings. When they lock in on an angle, the display will show the distance from the previous point and the angle as shown in Figure 3.3. If you have both Snap and Polar activated, then the point selected will be on the polar angle, even if that means pulling it off a snap point.

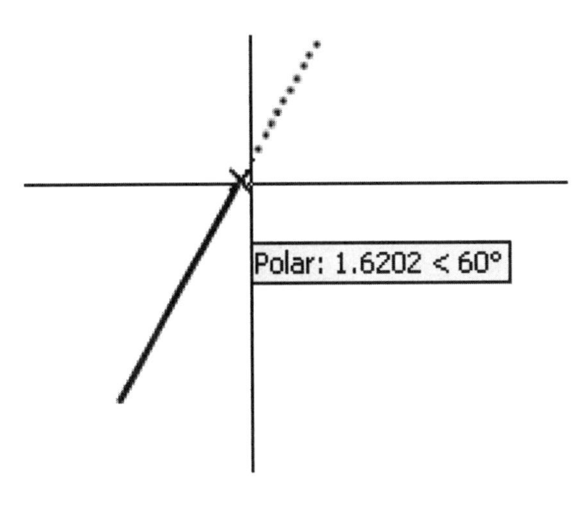

Figure 3.3

To activate Polar, pick the button labeled "Polar" at the bottom of the screen. This will activate Polar, and deactivate "Ortho" (if it was on). The angles will be set to whatever was last used on the computer. To set the angles, right click on the "Polar" button and select settings. The following dialog box will be displayed (Figure 3.4). To change the angle settings, drop down the "Increment angle:" box and select the new increment. In the box shown in Figure 3.4, the crosshairs will lock at 0°, 30°, 60°, 90°, etc. If you want to include any angles which are not multiples of the increment, they must be added to the list of additional angles.

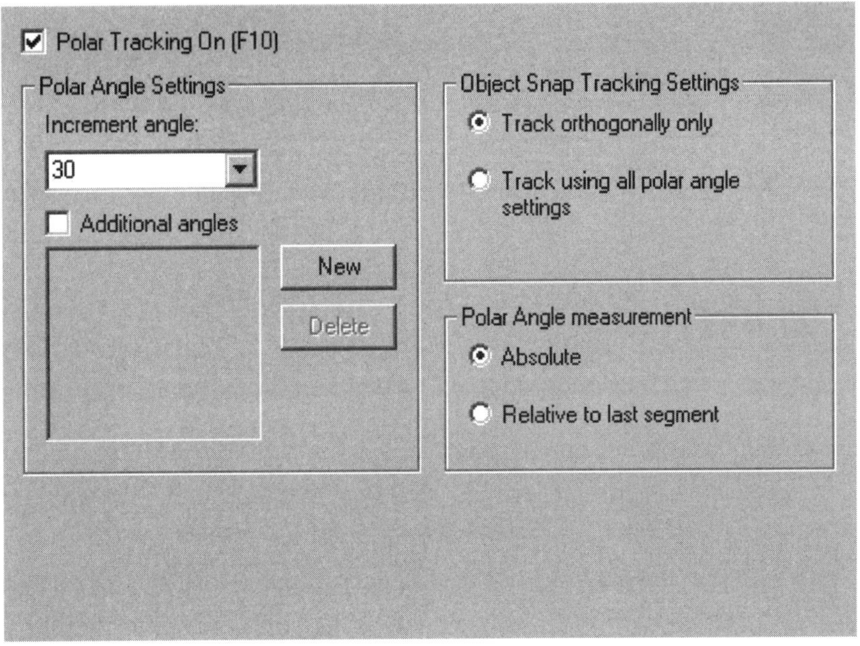

Figure 3.4

To add a 45° to the above setting, select the New button and type the additional angle on the list of angles. The new box should look like Figure 3.5.

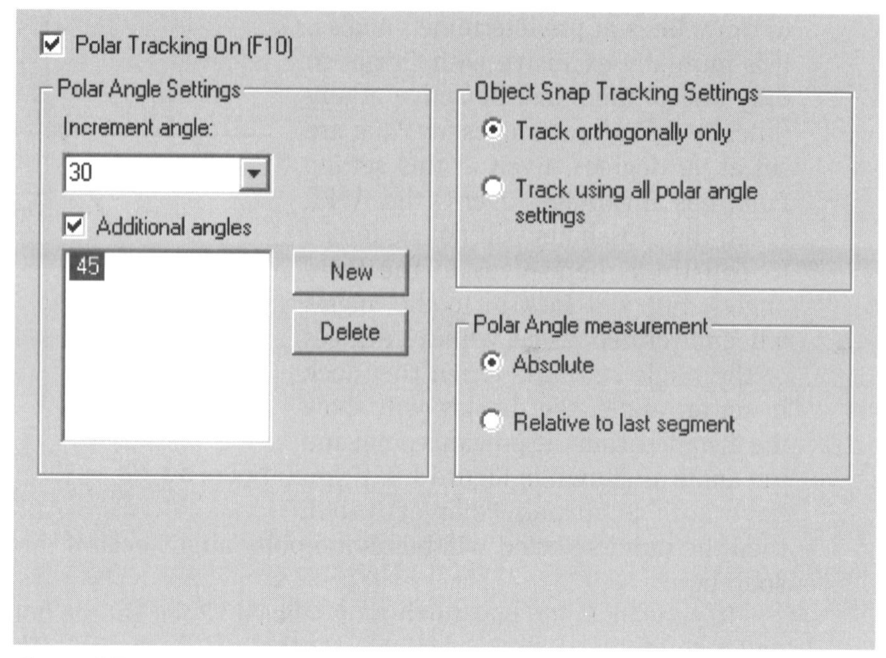

Figure 3.5

Which shows the additional angle of 45°. Note that this does not add 135°, 225°, or 315°. These would have to be added separately, if needed.

Resizing the Crosshairs

When AutoCAD is installed, the default size for the crosshairs covers only 5% of the screen. Many designers prefer crosshairs which span the entire screen. The size of the crosshairs is controlled by a setting called CURSORSIZE. This parameter represents the size of the crosshairs, in percent of screen size. To reset the size, simply type CURSORSIZE in response to the Command: prompt and type the desired value.

```
Command: cursorsize
Enter new value for CURSORSIZE <5>: 100
```

Use of Layers

While snap is one of the most important concepts while using a CAD system, layers are at about the same importance level. Using layers correctly will make it easier to work on large projects, especially those projects done by a team of engineers. Layers are a sorting system used by AutoCAD to keep track of objects and to categorize them based on the function they have within the drawing. Some lines are visible lines; some are construction lines. By identifying the function of the object to AutoCAD, it will correctly set the parameters associated with that type of line (i.e., don't print construction lines, print visible lines bold, draw hidden lines with dashes, etc.)

In this course, you should not have to create your own layers since the base templates (see later in this chapter) for the workbook have been created and are available on your computer in the classroom. For now, there are only two things you need to know about the mechanics of layers: how to switch from one to another and how to turn them on or off.

These two operations can be performed by using the dropdown layer selection box on the object properties toolbar shown in Figure 3.6 (the lower toolbar at the top of the screen). It will always display the name of a layer created in the current drawing if you are trying to draw an object. All objects must reside on a layer, they cannot be left "floating" and layerless. The layer on which they are drawn determines several properties of the object. These include: color, linetype, and lineweight. These parameters can be controlled individually for objects, regardless of layer, BUT THIS IS HIGHLY DISCOURAGED FOR BEGINNING STUDENTS. The three boxes to the right of layer name should be left as BYLAYER, thus allowing all objects to inherit their properties from the layer on which they are created.

Figure 3.6

To change from one layer to another you must drop down the list of layers and select a new one from the resulting list by clicking on the name of the layer. The layer visible will be the new current layer if the selection button was pressed while it is highlighted as shown in Figure 3.7.

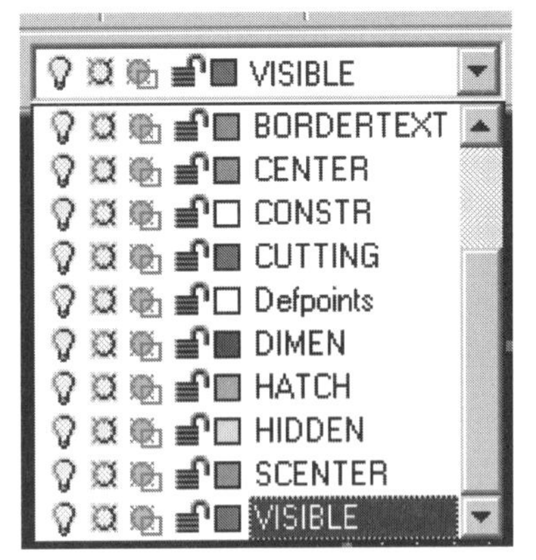

Figure 3.7

Limits

Limits are set in AutoCAD to allow the drawer to have a reserved area to draw in. They set the theoretical edges of the paper and help maintain perspective when drawing. The grid will only show within the limit defined in the drawing. To change the limits set for a drawing, use the command LIMITS and enter, via the keyboard, the lower left and upper right coordinates to define the page size.

Template Files

Template files are designed to allow companies or individuals to store predefined configurations for drawings. They preset many of the settings discussed in this chapter, thus allowing the CAD operator to begin drawing immediately, rather than take time to set up everything for the drawing. They can also include standard drawn objects, like a border or company logo.

For this class, template files allow your instructor to make copies of the basic setting needed to complete each drawing on your computer, thus relieving you of a lot of overhead required to start all drawings from scratch.

Object Snap

While snap is a very useful tool, there are times when you really need to draw to a point which is not on a snap point, and it might be very difficult to find exactly the snap you needed to force it on a snap point. One example is to draw a 3 point arc, then draw a line from the center of that arc to each of the end points.

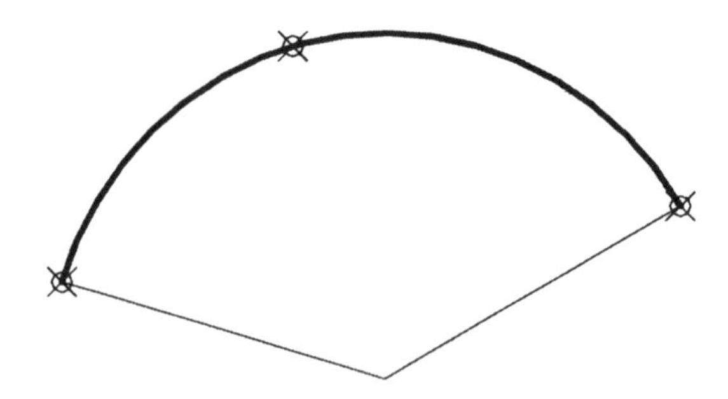

Figure 3.8

The chances that the center point will fall exactly on a snap point are slim to none. However, AutoCAD has an additional snap mode called object snap, OSNAP, to allow the user to lock onto points which are referenced by objects in the drawing by simply getting the crosshairs close to the desired locations. There are two different modes available for OSNAP, running and override. Running OSNAP is really a setting for AutoCAD, override is just a one point effect and it will be discussed only because it related so closely to running osnap.

To activate running osnap select the button labeled OSNAP at the bottom of the screen. The first time you do this you will be presented with a dialog box of available options as shown in Figure 3.9. Each subsequent time the button will either activate or deactivate the options you choose on this dialog box. If you want to see this box in the future, you must right click on the button and select settings.

Some of the various modes are:

Endpoint: This will choose the end points of an arc or line.

Midpoint: This will choose a point on an arc or line that is exactly in the middle. For an arc this is NOT the center point of the arc, it lies on the arc at its middle.

Center: The center of arcs or circles.

Quadrant: One of the points on an arc or circle which is at 3 o'clock, 6 o'clock, 9 o'clock, or 12 o'clock.

Intersection: The physical intersection between two objects. You may pick one object and then the second or locate the crosshairs over the intersection point and make your selection.

Insertion: The reference point for text.

Perpendicular: Selects a point on the object where a line perpendicular to the object and passing through the previous point (used in this command) would be. If this is the first point requested by the current command, it will allow a deferred perpendicular, which will not select the first point (the one on the object) until the next point is chosen. It then back calculates where the perpendicular needs to be and selects that point.

Tangent: Selects a point to allow the line, arc, or circle to be tangent to a selected arc or circle. Like the perpendicular option, this can be used in deferred mode to allow you to draw something tangent to an object and perpendicular to another.

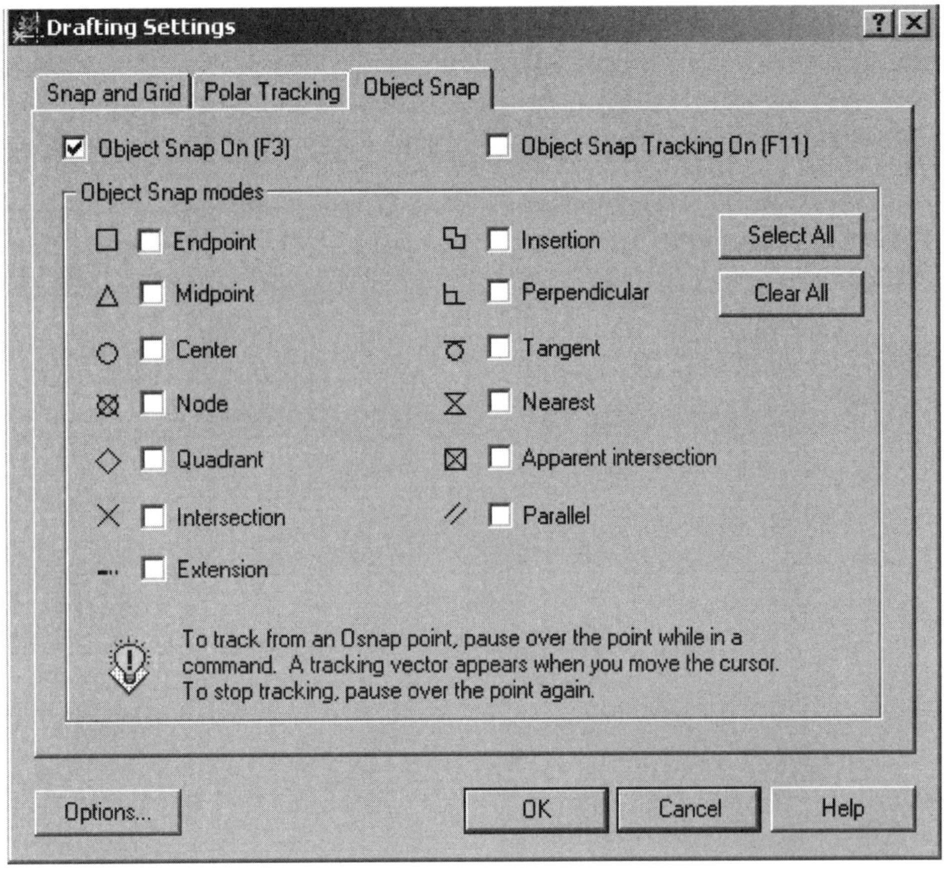

Drafting Settings
Snap and Grid
Polar Tracking
Object Snap
Object Snap On (F3)
Object Snap Tracking On (F11)
Object Snap modes
Endpoint
Midpoint
Center
Node
Quadrant
Intersection
Extension
Insertion
Perpendicular
Tangent
Nearest
Apparent intersection
Parallel
Select All
Clear All
To track from an Osnap point, pause over the point while in a command. A tracking vector appears when you move the cursor. To stop tracking, pause over the point again.
Options...
OK
Cancel
Help

Figure 3.9

Once you have selected the modes you would like to use, select the OK button and return to the drawing. The next time AutoCAD requests a point, it will try to observe the object snap rules you have established. If no point matches your osnap criteria, then it will select a point normally. If multiple points match the criteria, the AutoCAD will select the one closest to the actual crosshair location. If you want another possible selection, then pressing the tab key will cycle through all the available options.

The second mode, override osnap, functions exactly as running osnap except it is good for only one point selection, then the system reverts back to the running osnap settings. To use this option you may type the osnap mode in via the keyboard (the first three letters are sufficient) or select the icon from the dropdown toolbar icon shown in Figure 3.10. The one additional setting that is available for override osnap is "none" which cancels any running osnap modes for the next point selection.

Temporary Tracking
Snap from
Endpoint
Midpoint
Intersection
Apparent Intersection
Extension
Center
Quadrant
Tangent
Perpendicular
Parallel
Insert
Node
Nearest
None
Object Snap

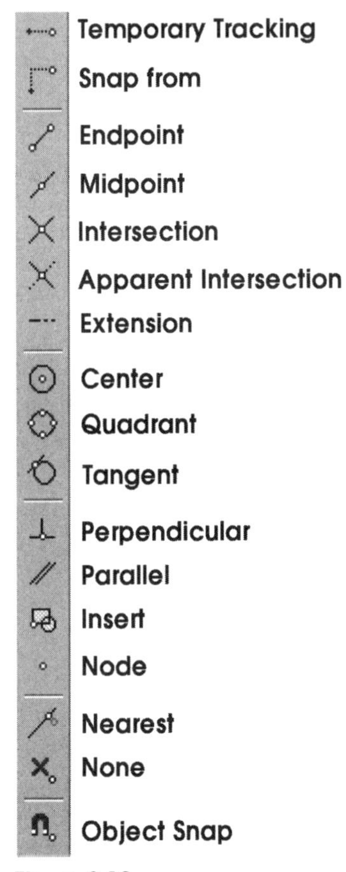

Figure 3.10

One closely related option is Object Snap Tracking. Beginning students should probably just make sure OTRACK (on the button bar at the bottom of the screen) is turned off. It will cause AutoCAD to create temporary tracking points using the OSNAP and polar angle snap settings. There are some very useful things which can be done with OTRACK, but it tends to be confusing to the beginning student.

Activating Toolbars

AutoCAD comes with 24 different toolbars available, fortunately they are not all displayed on the screen at the same time. If they were, the screen would look like Figure 3.11.

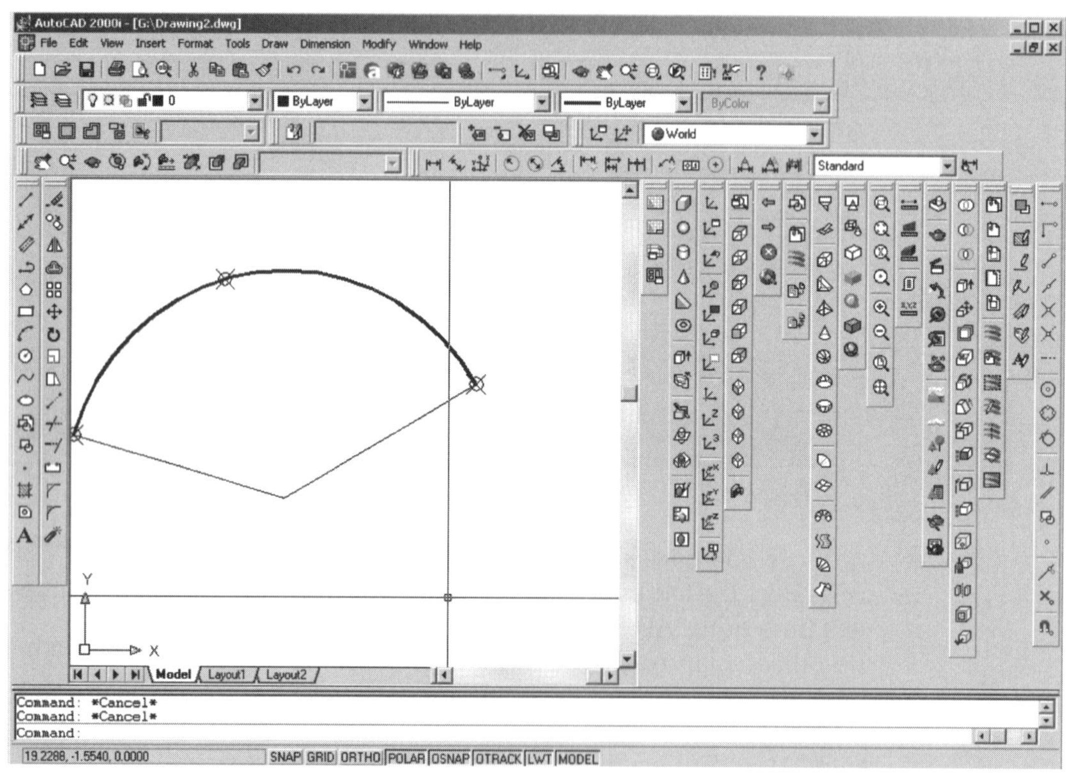

Figure 3.11

If you would like to open a toolbar which is not visible, you may right click on any existing toolbar and select the new toolbar from the resulting popup menu. If no toolbars are visible, then you can use the pulldown menu VIEW → Toolbars . . . to activate them. When AutoCAD first loads, the following toolbars are normally visible on the screen: "Standard Toolbar" and "Object Properties" at the top and Draw and Modify on the side. If, for some reason, you do not have these visible, you may reactivate them and drag them into place.

Display Control

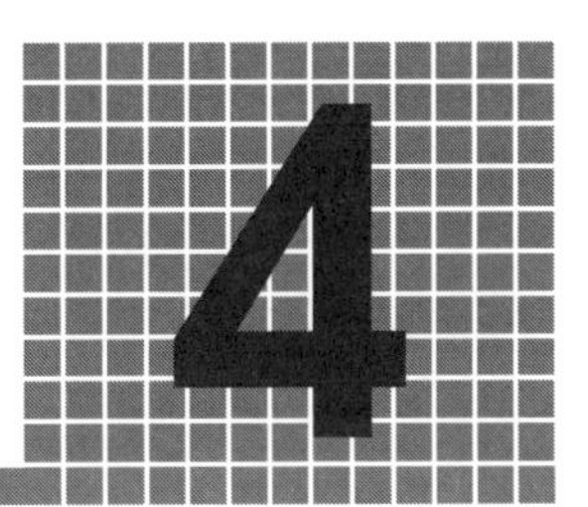

Objectives

1. Be able to ZOOM and PAN a drawing
2. Be able to use named views to display the correct portion of a drawing
3. Be able to change the visibility of a layer
4. Understand BLIPMODE
5. Refresh the display by using REDRAW or REGEN
6. Understand scaling of drawings
7. Use LTSCALE

Zoom and Pan

One of the advantages of a CAD system of traditional pencil and paper drawings is the ability to expand a portion of the drawing to make it easier to see what is being drawn. This feature is called zooming. AutoCAD has many options for the command ZOOM to allow you to display exactly the area of your drawing you desire. The keyboard command is ZOOM or just Z. The most common options are:

Window: The window option allows you to specify to points which form diagonally opposite corners of a rectangle. The portion of the drawing contained within that rectangle will be expanded to fill the screen, however, AutoCAD will not change the proportions of that which is displayed, it will show more of the drawing in one direction or another to keep it proportional. For instance, if the drawing on the left in Figure 4.1 were to have the ZOOM:Window performed, the result will be the drawing on the right.

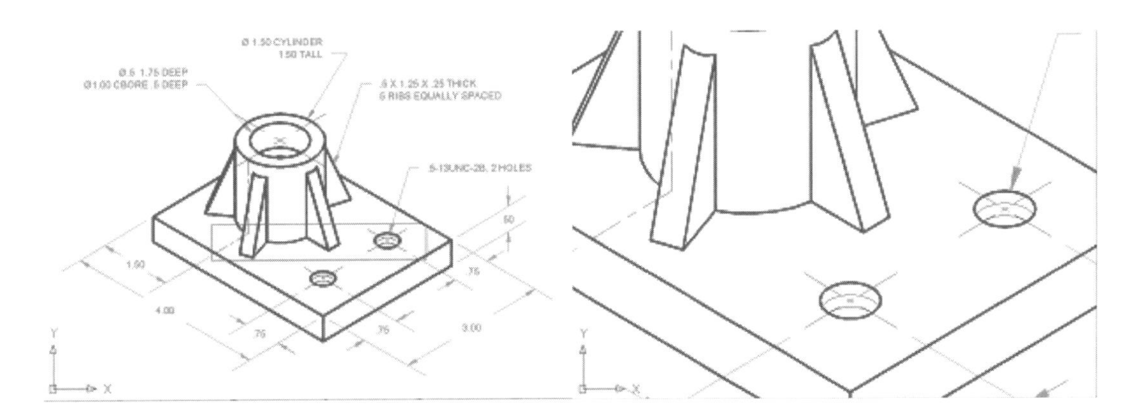

Figure 4.1

Extents: Once you have zoomed in, you may return to see the entire drawing by using the ZOOM:Extents option. This will fill the screen with everything you have drawn (and is on a thawed layer).

All: This option will fill the screen with all the objects you have drawn (like extents), but it will also insure that the entire Limits that are defined for the drawing are visible on screen.

Realtime: This will allow you to dynamically change the display of your drawing by zooming in (making it larger) or zooming out of the drawing by holding the left mouse button and moving the mouse up or down (side to side movement has no effect). By moving the mouse up, you cause the picture to get larger. The cursor will shift to a magnifying glass with a small + and – beside it.

Previous: AutoCAD stores the previous displays and will return to them, in sequence, by using ZOOM:Previous. AutoCAD will store 10 previous views.

In addition to these options, if your system is equipped with an IntelliMouse or other mouse which has a roller wheel between the right and left buttons, you may roll the wheel to zoom in or out.

Unlike Zoom, PAN does not magnify the drawing, but allows you to slide the drawing under the monitor to display parts of the drawing which are currently off screen. Pan has no options, when you select the icon (or use PAN or just P through the keyboard) the cursor becomes a hand and by holding down the left mouse button, you can drag the contents of the window.

The IntelliMouse wheel can also be used to pan your drawing. By holding down on the wheel and moving the mouse you can drag the contents as with panning.

When using ZOOM:Previous to restore the previous view, changes via the PAN command count as views saved, so you can ZOOM:Previous to return to the view prior to panning the drawing. One limitation to using ZOOM:Previous is that the previous views are not saved with the drawing, so when you begin a new drawing session all previous zooms are lost.

Using Named Views

Sometimes you would like to be able to return to a particular zoomed area of the drawing without having to go through all the intermediate zooms you have performed. If you recognize that a display is important, you can save the Zoom/Pan information in a Named View. Using Named Views allows you to return to those saved displays directly, without having to zoom previous multiple times. To save a view you use the VIEW command or select the icon. This will display the view creation dialog box shown in Figure 4.2. Using this box you can create a New saved view. The New View dialog box will be displayed and you will be asked for a view name. If you have already zoomed in on the area of the drawing you want associated with the given name, then all you need to do is select OK on the New View dialog box. This view name will now appear in the list of views below the view named Current.

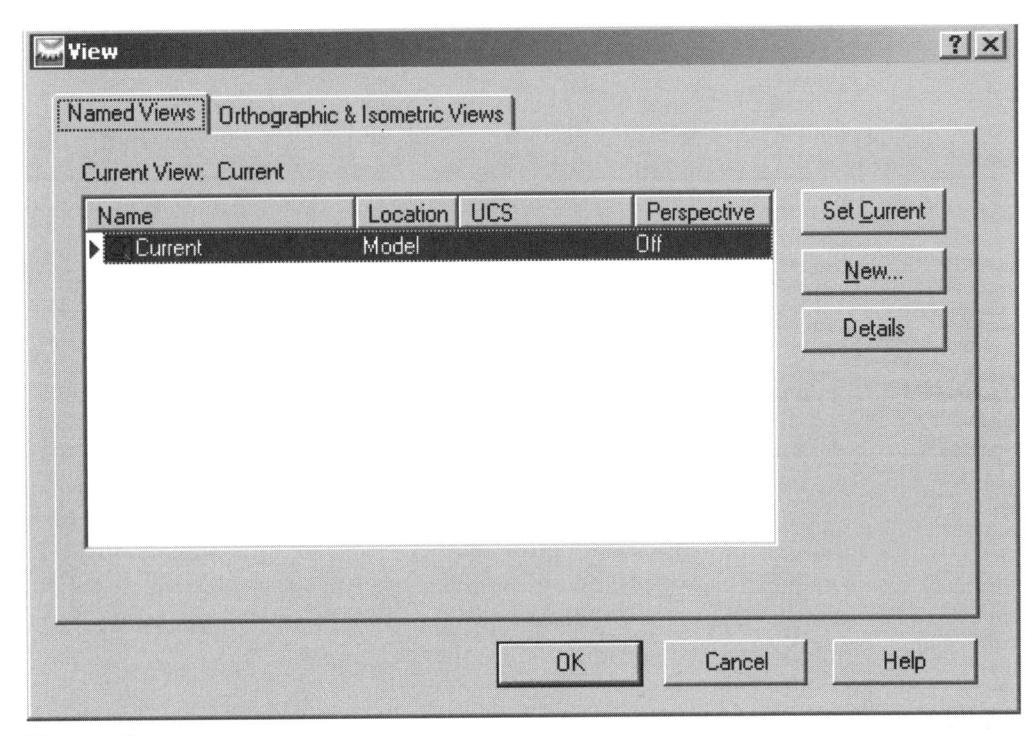

Figure 4.2

Many of the items requested with this command are more suited to 3D drawings and we will come back and revisit the options later. For now, do not worry about the "Orthographic & Isometric Views" tab as these are strictly 3D options.

If you have saved a view previously, you may restore it to the display by selecting the view from the list and selecting the Set Current button.

Controlling Layers

While zooming to display different parts of the drawing is important, another area of display control is working with layers. In the previous chapter, we talked about changing from one layer to another and why this could be useful for organizing your drawing. One of the other advantages of layers is the ability to independently control which layers of a drawing are displayed at any given time. You can sort of think of layers as drawings on sheets of clear plastic, which can be overlaid on each other.

To change the visibility of a layer, drop down the layer list (Figure 4.3) and click on the light bulb to the far left of the layer name. If the bulb is bright (yellow), then the objects on that layer will be displayed, if it is dark (gray) then they will not. In the example shown in Figure 4.3, the layers Border, Constr, and Window will not be displayed on the drawing.

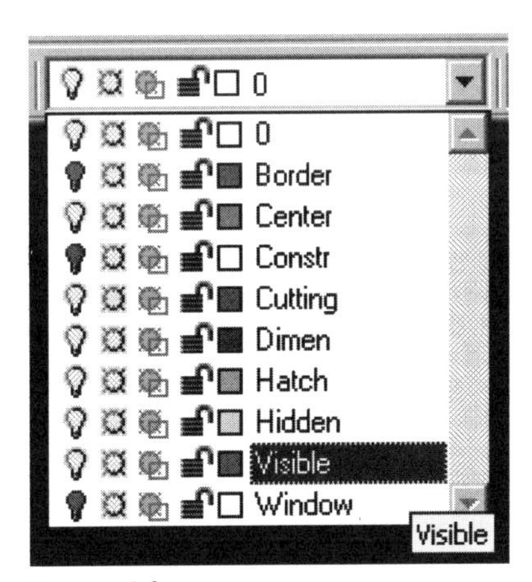

Figure 4.3

Blips

One of the features of AutoCAD which can be enabled are Blips. Blips are simply small marks put on a drawing to indicate where the crosshairs were when the select button was pressed. These are temporary marks just placed on the display for the benefit of the drawer. If you do anything to redisplay the drawing on the screen, the old blips will disappear. This could be a Zoom, Pan or simply turning the grid off.

If you would like to see the Blips, use the command BLIPMODE. Simply turn BLIPMODE on and the blips will begin to appear, turn it off to stop them from appearing.

Redraw or Regen

One other method to clear the Blips from the screen is to force AutoCAD to recalculate the proper display of the drawing on the screen. The two commands to accomplish this are REDRAW and REGEN. The effect on the screen is virtually identical, but the calculations behind the two commands differ. On a very complex drawing the REDRAW command is slightly faster, but in some 3D applications it is less effective. The REDRAW command can be abbreviated by "R". For the beginning student the two commands are interchangeable.

Scaling of Drawings

Many times a drawing will not fit on the paper at full scale, or it may appear too small. If this is the case, some basic changes need to be made to the operation of AutoCAD to allow these drawings to be completed. The information needed is simply the necessary scale to complete the drawing. This can be calculated by dividing by page size by the size of the object to be drawn (remember to account for multiple views and the space between them, if needed). Generally, the view spacing (and therefore, the space needed to dimension the views) is deducted from the page size along with the edge of the page margins, and this "effective page size" is divided by the sum of the object dimensions to arrive at a scale factor.

Once that scale is established, you have to shift your entire mind set from the method of drawing on paper to a new method for CAD. On paper you adjust the size of your object to fit on the page, in CAD you adjust the size of your page to fit the drawing. This is a necessity due to the first cardinal rule of Computer Drafting:

ALWAYS DRAW OBJECTS FULL SIZE IN THE COMPUTER

This allows different draftsmen to work on projects and still share files without having to remember what scale everyone was working at.

Assuming that you are working from a basic template you should make the following changes BEFORE you begin your solution:

You must adjust the size of the border to account for the different page size. If you would have to reduce the size of the object to fit on the page, you must instead increase the size of the paper to accommodate the object. Thus if you were trying to draw a top view of a Boeing airplane which had to be scaled at 1 = 100, you would need to increase your border size by a factor of 100. This is done by using the SCALE command. You select the entire border and list the base point as 0,0. This will cause the border to stay in the upper right quadrant. The scale will be 100 in this case. After this scale the border will be well off screen, so you should do a zoom all to see the new border at its new size.

In order to get the grid to appear you will need to adjust the size of the grid by about the same factor. Since the default grid in the workbook templates is 0.2", you would need to change the grid to about 20". In the command stream shown below I will use a value of 24", thus having a 2 foot grid. However, when you change this, the grid does not fill the border. Why?

The why deals with Limits. The grid will only fill the limits set for the drawing, and scaling the border does not reset the limits. This must be done with the LIMITS command. Both the lower left corner and the upper right corner values need to be multiplied by 100.

You will probably also need to adjust your snap to a convenient value based on what you are drawing and the precision needed there.

Other changes are in your mind set. If you want to leave 1" between the top and front views of the plane, then you will have to have a 100" gap in the computer, since everything will be reduced by a factor of 100 when it is eventually sent to hard copy. Likewise, your text height will have to be increased by a factor of 100 to be readable. (Think about taking your pencil out to the plane and printing in nice 1/8" single stroke gothic lettering the plane's serial number, then stepping back and taking a photo of the plane and printing that on a 8" x 10" picture. Do you think you could read what you wrote?)

The last change to discuss now (more changes will be necessary when we discuss dimensioning in Chapter 8) deals with dashed lines. Hidden and center lines will need to be adjusted just like the text height would. (Once again, go back to the concept of drawing a center line on the side of the plane and photographing it.) There is an overall scale factor that will adjust the size of all dashed lines in a drawing. It is called LTSCALE. By setting it to 100, all dashes and gaps in all dashed lines will be 100 times larger.

```
Command: scale
Select objects: {put a window around the border}
Specify opposite corner: 102 found
Select objects:
Specify base point: 0,0
Specify scale factor or [Reference]: 100
Command: zoom
Specify corner of window, enter a scale factor (nX or nXP), or
      [All/Center/Dynamic/Extents/Previous/Scale/Window] <real
      time>: all
Regenerating model.
Grid too dense to display
Command: grid
Specify grid spacing(X) or [ON/OFF/Snap/Aspect] <0.2500>: 24
Command: snap
Specify snap spacing or [ON/OFF/Aspect/Rotate/Style/Type]
      <0.1250>: 12
Command: limits
Reset Model space limits:
Specify lower left corner or [ON/OFF] <0.0000,0.0000>: 0,0
Specify upper right corner <10.0000,7.5000>: 1000,750
Command: ltscale
Enter new linetype scale factor <1.0000>: 100
Regenerating model.
```

Paperspace, Modelspace, and Printing

Objectives

1. Be able to print a drawing
2. Understand Layout mode and Model mode

Printing Drawings

So far, we have learned the basics of creating drawings and it is time to discuss sending them to a printer. The full process of printing drawings in AutoCAD can be a bit intimidating, but once you see the rationale behind all the options and get some of the basic settings set up, the process can be streamlined greatly.

It all begins with the command PLOT (which comes from a time when pen plotters were the preferred method of drawing output). If you use the command PRINT, you will notice that it is aliased to the PLOT command. Doing this will launch the initial dialog box for printing shown in Figure 5.1.

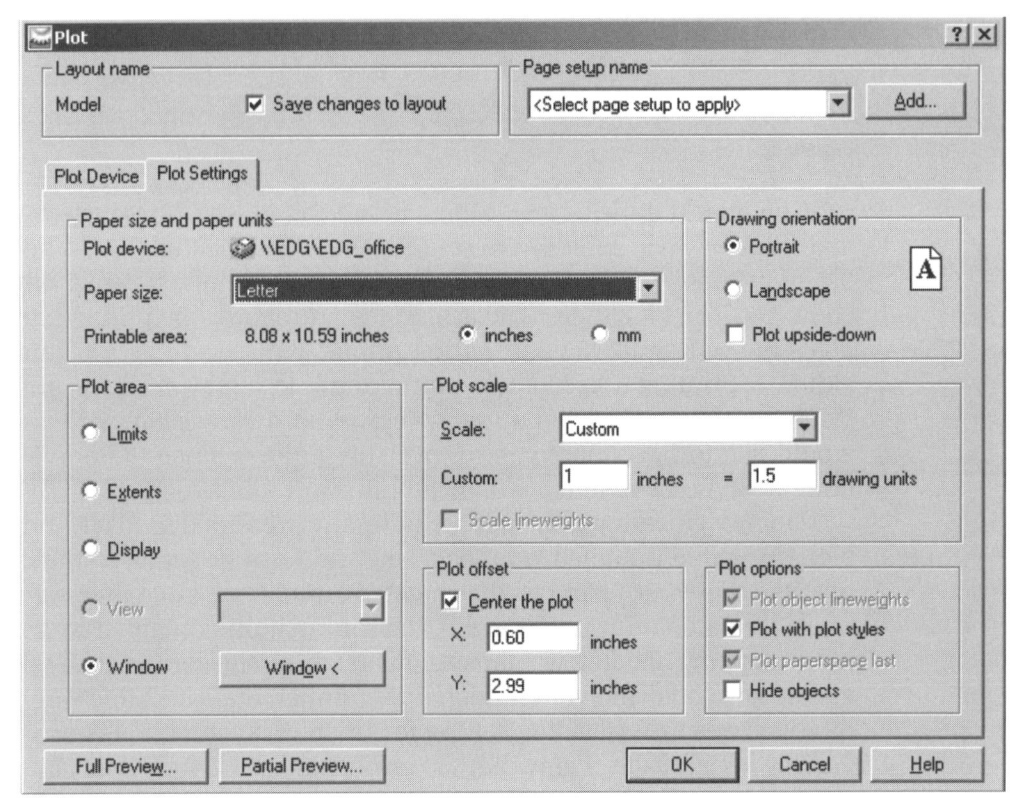

Figure 5.1

There are two tabs available: Plot Device and Plot Settings. Both are important to getting correct output from AutoCAD. We will start by looking at the Plot Device tab, since we must tell AutoCAD what printer to use (it is also possible to plot the drawing to a disk file. We will discuss how to do this in chapter 12). This tab is shown in Figure 5.2.

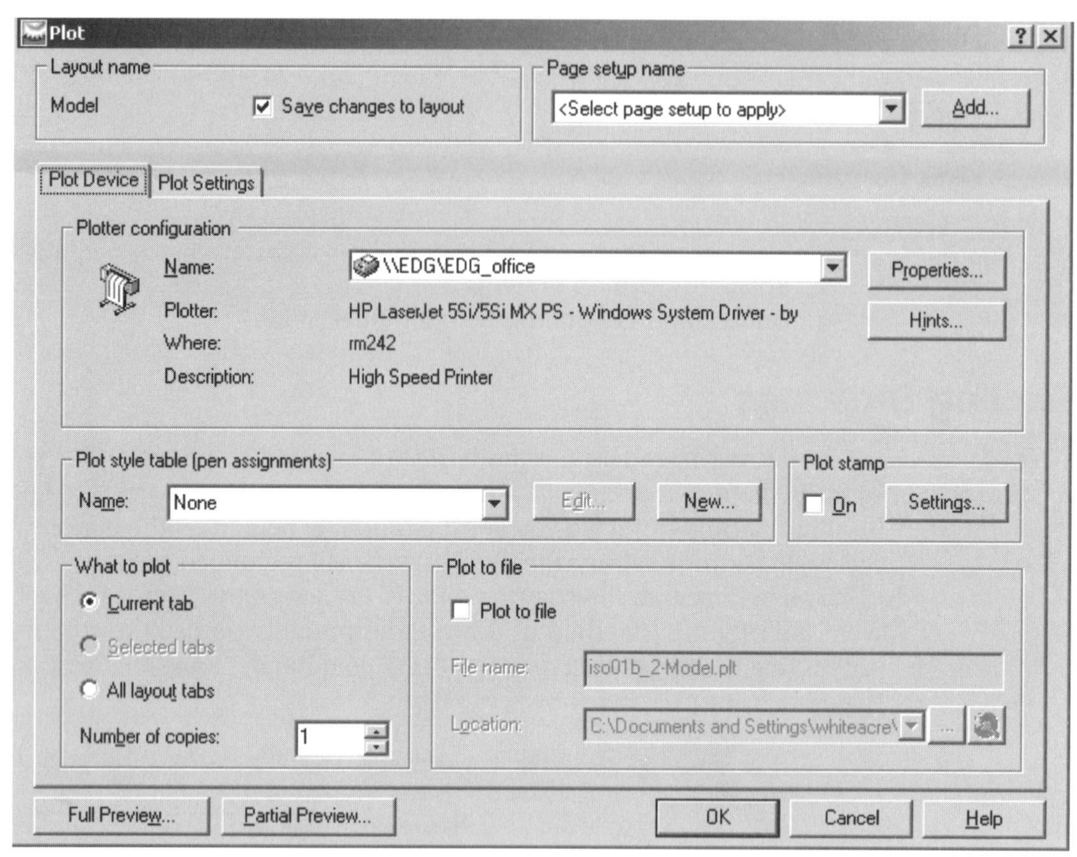

Figure 5.2

The important features of this tab are the Plotter Name (here set to EDG_office) and the Plot Style Table (here set to None). The printer should be selected from the list of available printers on your computer. It simply controls where the print will come out. The Plot Style Table is used to control how the printer handles colors. If you leave the setting to None, then the printer will decide how it interprets color data, either by printing it in full color, dithering it to a gray scale, or printing in Black and White. Leaving a decision up to a piece of hardware is generally a bad idea, and you would like to have more control over the process since there is more which can be done with a plot style table than simply controlling color.

Since engineering drawings are normally presented in Black and White, the logical plot style table is named "monochrome" and can be found in the dropdown list. It is possible that your instructor has created a plot style table precisely for this course. If so, make sure you use that style. The information contained in a plot style includes color mapping and line weight specifications, so the correct style table will make your visible lines bold and your hidden lines thinner. If all your lines are printing very thin, then you need to check the setting for the plot style table. Likewise, if your lines are varying in the shade of gray they are printed in, check your style table setting. Normally this need only be done once per drawing, since the settings stay with the drawing, not the computer. The template file can even have this information embedded in it.

Once you have those two settings done, you should return to the Plot Setting tab to finish the preparations for printing. On this tab you should check, in order:

Paper Size and Units: These should be set to the correct size for your page (normally Letter) and the correct set of units (inches or millimeters) depending on what system was used to construct your drawing.

Drawing Orientation: Either Portrait or Landscape depending on the setup of your drawing.

Plot Area: Your choices are: Layout (or Limits), Extents, View, or Window. Some of these may be grayed out if they are not appropriate for the given conditions. This is probably the second most misapplied section when plotting. You must know what it is you WANT to plot and then select the option that corresponds to your needs. Layout will plot all the drawing that falls on the theoretical page shown in the layout mode (this is discussed later during the section on 3D). It is only an option if you are in layout mode. If you are in model mode, then the option will be Limits and this will plot only that portion of your drawing which exists within the limits set for your drawing. Extents will plot your entire drawing, from top to bottom and left to right. View will plot whatever is saved for that named view. Window will be available only after you have selected a window on your drawing to print. This is done by selecting the "WINDOW" button and picking diagonally opposite corners as if you were zooming in on the area.

Plot Scale: This determines the relationship between one unit in the drawing and one unit on the printed page. The printed page units are either inches or millimeters as set under the Paper Size and Units area. The scale is the most commonly misused setting when plotting. As a general rule, scaling your drawing to "Fit" is a *bad* option. It tells AutoCAD to calculate whatever scale factor is needed to get the Plot Area to fit on the Paper Size. You normally want to have the output printed at a specified scale. For a test print just to allow you to mark it up and make corrections this is not as important, but for work which is to be submitted it is critical. You can select a scale which prints your entire Plot Area on a small portion of the page, or you can select a scale which makes the Plot Area larger than the paper available. If the Plot Area is too large for the printed page, it will be clipped at the edge of the paper and AutoCAD will NOT issue any warnings.

Plot Offset: This determines where on the page the lower left corner of the plot area will be located. It is in Page Size Units. Choosing the Center Plot option will place the center of the Plot Area at the center of the paper.

Plot Options: You normally want to "Plot with Plot Styles" since this is a good way to control the appearance of your output. Plotting paperspace last does not really make any difference for printers, however if you are using a pen plotter, it is good to print all of the drawing first and the border last. Hiding objects is a 3D option and will generally be left unchecked even in 3D.

Once you have all the options set the way you think you want them, do a "Full Preview" and see what is displayed as being the plotted output. AutoCAD will generate precisely what your output will look like on the printed page. The bottom line is: *If the print preview does not look right, the hard copy won't look right either!* Save a few trees and always do a Full Preview before sending the drawing to the printer.

Using Layout Mode

Many first year courses do not plot the 2D drawings using Layout Mode. If your drawing on the Model Tab has a border and title information, then you do not need to use Layout mode until you begin constructing drawings in 3D.

To facilitate laying out of drawing for printing, AutoCAD uses a special system called LAYOUT mode. When you create a new drawing from scratch, you will be able to see that there are three tabs near the lower left side of the drawing labeled Model, Layout1, and Layout2. When you select one of the layout modes for the first time, you will be presented with a dialog box asking about the configuration of the page you want to use for this layout. This box is very similar to the plotting dialog box and the parameters should be set as in the plot dialog boxes above. What will then be displayed is a special area of the drawing called Paperspace while your drawing is done in Modelspace.

The only reason for mentioning it here is the inevitable chance that you clicked the button at the bottom of the screen labeled "Model" and you were taken directly to Layout mode and Paper Space (as indicated by the background changing from black to white). Unfortunately, just clicking on the button (now labeled Paper) will not get you back to where you were. If this happens, all you have to do is select the tab labeled Model and you will be restored to normal model space and drawing mode.

When we begin 3D drawings, we will have much more to say about Layout mode.

Editing, The True Power of CAD

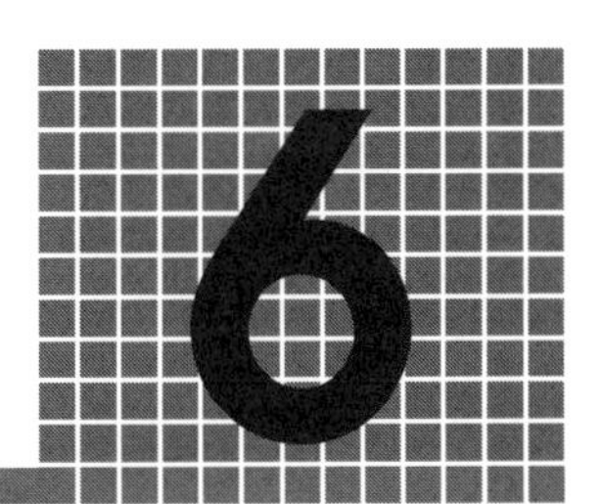

Objectives

1. Be able to create selection sets in AutoCAD
2. To be able to edit your drawings in AutoCAD
3. The ability to use keyboard coordinates
4. To be able to use Grips to edit drawings in real time
5. To be able to change the layer of an object already created
6. Edit already created text
7. Modify the basic properties of an object

By now, you have the ability to create some fairly complex drawings in AutoCAD. You can draw lines, circles, arcs, and write text. You can plot your drawings and get high quality output from the computer. All these things are good, however they are just the tip of the iceberg of CAD. Using the tools currently at your disposal, you will not be much more productive than a traditional hand draftsman. This chapter will open the doors to increasing your productivity and making CAD pay for itself.

As the chapter title says, the true power of CAD does not lie in the creating of lines, but in the ability to edit them. It lies in the ability to draw something once and never have to create that part again, to be able to copy that part to wherever it is needed. It lies in the ability to create construction lines which can be perfectly erased. It lies in the ability to edit a drawing.

Selection Sets

We have actually already used one editing command, ERASE. Using that command we can begin to see the applications of selection sets. Selection Sets are the beginning of any editing command. They determine what objects are to be edited. For the ERASE command they determine what objects will be deleted from the drawing and which will remain. In all editing commands there is a need to specify precisely which objects will be the target of the edit and this is what selection sets do.

A selection set is formed in response to the prompt "Select Objects:" in AutoCAD. From this prompt you will select the objects you want to be the target of the editing command. You do not have to simply select the objects one at a time, there are many options available for selecting the objects in groups. The simplest two are Windowing and Crossing as described earlier. Three additional options which do not require using the mouse are:

Last: this selects the last object drawn in the drawing

Previous: this re-selects the last selection set operated on

All: this selects all the objects in the drawing

Using those 6 options for selection sets will allow you to make use of the full editing capabilities of AutoCAD.

The editing commands available on the Modify toolbar are shown in Figure 6.1.

Erase
Copy
Mirror
Offset
Array
Move
Rotate
Scale
Stretch
Lengthen
Trim
Extend
Break at a point
Break
Chamfer
Fillet
Explode

Figure 6.1

Move and Copy

The MOVE and COPY commands function exactly the same, except the COPY command leaves a copy of the selected objects behind. They begin like all the other editing commands by requesting what you want to move or copy. You must form a selection set containing the objects you want and then press enter to terminate the selection set process.

You will be asked for two points, specifying a displacement vector over which the objects will be translated. Once you select the second point, the command will be executed and you will be returned to the Command: prompt.

```
Command: copy

Select objects: {pick those objects you want
      to move}
Specify opposite corner: 33 found

Select objects: {press Enter when you have completed your
      selections}
Specify base point or displacement, or [Multiple]: {pick the
      beginning point for the copy vector}
Specify second point of displacement or <use first point as
      displacement>: {drag the copies to the correct location}
```

Keyboard Coordinates

In addition to using the mouse to specify points in AutoCAD, it is possible to type a location in directly via the keyboard. This is called using keyboard coordinates. There are two formats of keyboard coordinates: Absolute or Relative. An absolute coordinate references the origin of the drawing, while a relative one references the last point selected. Absolute is the default, to note a relative coordinate place an "@" before the coordinate values.

There are also 2 styles of keyboard coordinates: Cartesian and polar.

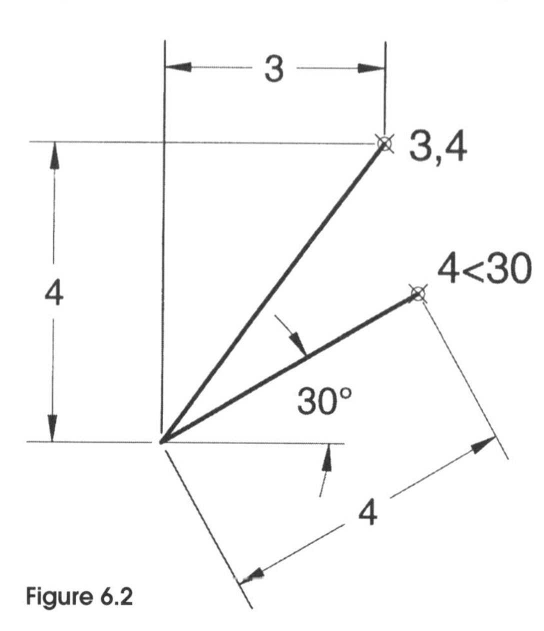

Figure 6.2

Cartesian coordinates reference X and Y, while polar reference a distance and an angle from the horizontal. For Cartesian, the format is simply "X,Y" and for polar it is "dist < angle".

You can mix either format with either style to allow for four different input formats. These are:

Absolute Cartesian: 3,4
Relative Cartesian: @3,4
Absolute Polar: 4 < 30
Relative Polar: @4 < 30

Another option for specifying a point via the keyboard is the Direct Distance option. If AutoCAD is requesting a point, you may type a single number in through the keyboard. AutoCAD will measure the angle from the last point selected to the current locations of the crosshairs and will select a point as if these were the result of a relative polar input using the specified distance in the direction of the crosshairs.

Using keyboard coordinates along with the copy command is a good way to generate oblique pictorial drawings.

Break, Trim, and Extend

These three commands form a loosely related set. Break and Trim allow you to remove a portion of an object without erasing the entire object and Extend is the inverse operation for Trim.

To break an object, simply select the object at one side of the break and select the other end of the break for the second point. The "break at a point" icon simply uses the selected point as both the first and last point, thus splitting a line or arc into two distinct pieces.

```
Command: break
Select object:
Specify second break point or [First point]:
```

The trim command will remove a portion of an object like the break command, but it requires the existence of objects to form a boundary for the trimming operation. If you wanted to remove the portion of the line which is inside the circle as shown in Figure 6.3, you would have the circle be your cutting edge and the central portion of the line as the object for trim.

```
Command: trim
Current settings: Projection=UCS,
      Edge=None
Select cutting edges ...
Select objects: {pick circle} 1
      found

Select objects:
Select object to trim or shift-
      select to extend or [Project/
      Edge/Undo]: {pick center of
      line}
Select object to trim or shift-
      select to extend or [Project/
      Edge/Undo]:
```

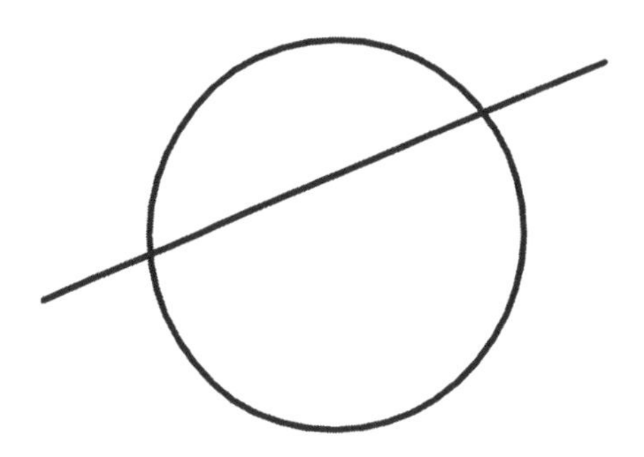

Figure 6.3

If you do not choose any cutting edges (i.e. just press enter at that prompt), then AutoCAD will act as if you had selected the entire drawing as cutting edges, thus any item to trim will be trimmed to the nearest intersection.

The extend command works just backward from the trim command. You must still specify boundaries, but these are the boundaries to which the objects will be extended.

Notice that you can extend objects using the TRIM command by holding the shift key when selecting the object to extend.

Example:

Draw a Cavalier Oblique of the object shown in Figure 6.4.

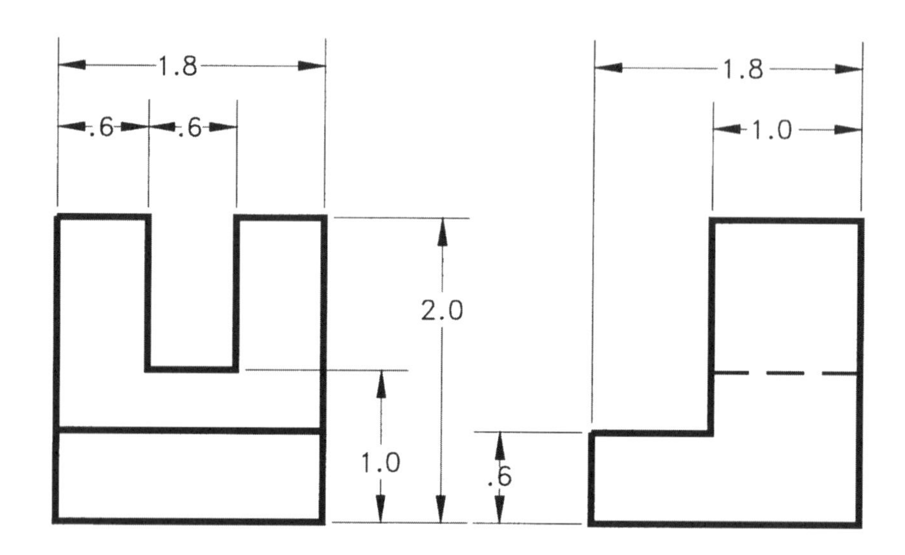

Figure 6.4

1. Draw the front view using normal AutoCAD commands and techniques.
2. Use the COPY command to replicate the front view to depict the depth.

```
Command: COPY
Select objects: Other corner: 11 found
Select objects:
<Base point or displacement>/Multiple: m
Base point:
Second point of displacement: @.8<40
Second point of displacement: @1.8<40
Second point of displacement:
```

3. ERASE the portions of the depth profiles which will not be seen.

```
Command: erase
Select objects:
Select objects:
...
Select objects:
Select objects:
```

4. Add receding lines using OSNAP and the LINE command.

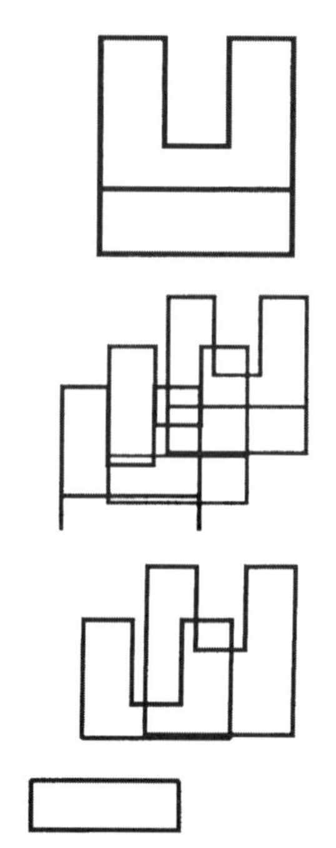

Figure 6.5

```
Command: line
From Point: endp of
To Point: endp of
To Point:
...
```

5. TRIM and ERASE the remaining lines which would not be visible.

```
Command: TRIM
Select cutting edges: (Projmode = UCS, Edgemode = No extend)
Select objects: 1 found
Select objects:
<Select object to Trim>/Project/Edge/Undo:
<Select object to trim>/Project/Edge/Undo:
```

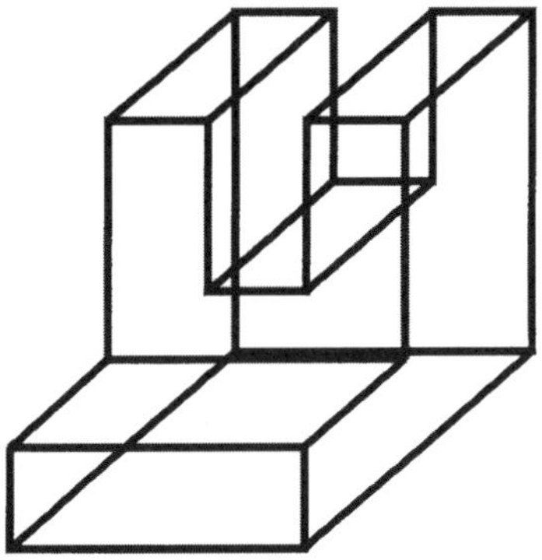

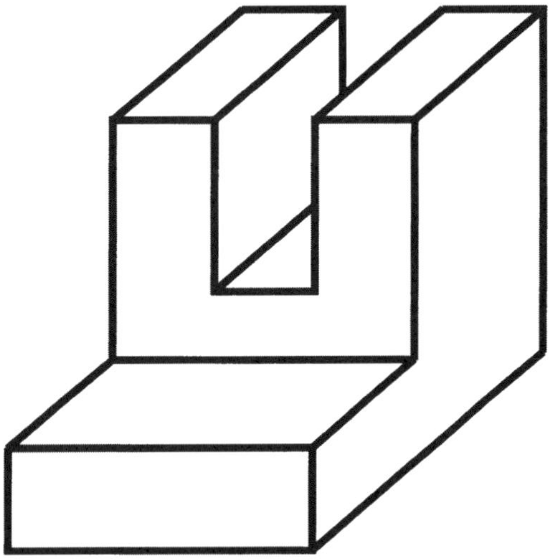

Figure 6.6

Chamfer and Fillet

Both these commands join existing lines or arcs. The fillet command will lay in a radius to complete the intersection while the Chamfer will extend the objects until they intersect and then cut a sharp corner from that intersection. In each case, the lines or arcs will be extended or trimmed as needed to reach the point of intersection prior to filleting or chamfering.

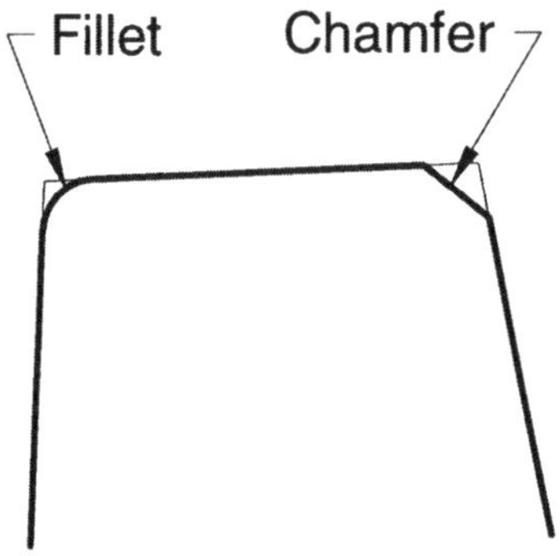

Fillet
Chamfer

Figure 6.7

```
Command: fillet

Current settings: Mode = TRIM, Radius =
      0.2500
Select first object or [Polyline/
      Radius/Trim]: R
Specify fillet radius <0.2500>: .5

Select first object or [Polyline/Radius/Trim]:
Select second object:
Command: chamfer
```

```
(TRIM mode) Current chamfer Dist1 = 0.5000, Dist2 = 0.5000
Select first line or [Polyline/Distance/Angle/Trim/Method]: T

Enter Trim mode option [Trim/No trim] <Trim>:

Select first line or [Polyline/Distance/Angle/Trim/Method]: M

Enter trim method [Distance/Angle] <Distance>:

Select first line or [Polyline/Distance/Angle/Trim/Method]:
Select second line:
```

Notice that you must specifically alter the fillet radius (or chamfer distances) prior to selecting the objects to be filleted. The Trim option will allow you to not have the objects extended to their intersection before applying a fillet or chamfer.

Other Editing Commands

Mirror

The mirror command will create a mirror image of the selected objects. You must select the objects to be mirrored and then select two points to define the line about which they will be mirrored. You can choose to retain or delete the original objects.

```
Command: mirror
Select objects: Specify opposite corner: 14 found

Select objects:
Specify first point of mirror line:
Specify second point of mirror line:
Delete source objects? [Yes/No] <N>:
```

Offset

Offset will create a copy of an object in a parallel fashion. For lines it will create a parallel line, for arcs or circles it will create a concentric arc or circle. You may either specify the distance between the objects or a point through which the offset copy must pass.

```
Command: offset
Specify offset distance or [Through] <1.0000>: .50

Select object to offset or <exit>:
Specify point on side to offset:
Select object to offset or <exit>:
```

Scale

Scale will allow you to increase or decrease the physical size of objects in your selection set. You must specify a base point about which all the scaling will occur.

```
Command: scale
Select objects: Specify opposite corner: 15 found
Select objects:
Specify base point:
Specify scale factor or [Reference]: .5
```

Stretch

Stretch will allow you to move one end of a line or arc without moving the other. The resulting object will be longer or shorter than the original. The prompts are essentially the same as for the move command.

```
Command: stretch
Select objects to stretch by crossing-window or crossing-
      polygon...
Select objects: Specify opposite corner: 2 found

Select objects:
Specify base point or displacement:
Specify second point of displacement or <use first point as
      displacement>:
```

It is worth observing the statement that selections must be made with a crossing window or crossing polygon. For this command you cannot just select an object and go, you must use a crossing selection and thereby specify not only the object, but which end of it you want to move.

Lengthen

Lengthen will modify the length of a line or arc. You can change the total length, the amount to be added (or subtracted), a percentage change, or, for arcs, the included angle can be modified.

```
Command: lengthen

Select an object or [DElta/Percent/Total/DYnamic]:
Current length: 2.0507
Select an object or [DElta/Percent/Total/DYnamic]: T
Specify total length or [Angle] <1.0000)>: 2.25

Select an object to change or [Undo]:
Select an object to change or [Undo]:
```

Array

Array is used to make many copies of a set of objects in a regular pattern, either rectangular or circular. A rectangular array makes copies of objects in rows and columns, similar to the arrangement of the desks in a classroom or windows on a high-rise building. A circular (actually called polar by AutoCAD) makes copies equally spaced along an arc or circle, like the spokes on a wheel or teeth on a gear.

For a rectangular array you must specify the number of columns and the number of rows. Columns are vertical lines offset in the "X" direction, rows are horizontal lines offset in the "Y" direction. Once you have specified how many columns and rows you want, you must specify the offset in both "X" and "Y." This is the distance from one point on the objects to the identical point on the next copy. Figure 6.8 shows an array of 3 rows and 5 columns of a diameter 1 circle, with row and column spacing of 0.75 in both directions.

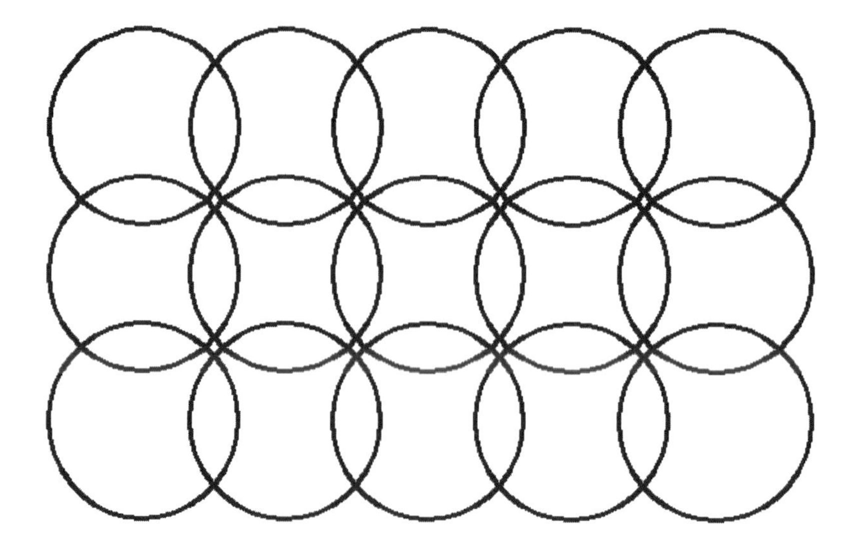

Figure 6.8

To create this pattern draw a diameter 1.00" circle and then use the array command. It will display the following dialog box (Figure 6.9). Complete the box as shown and select the circle under the "Select objects" button.

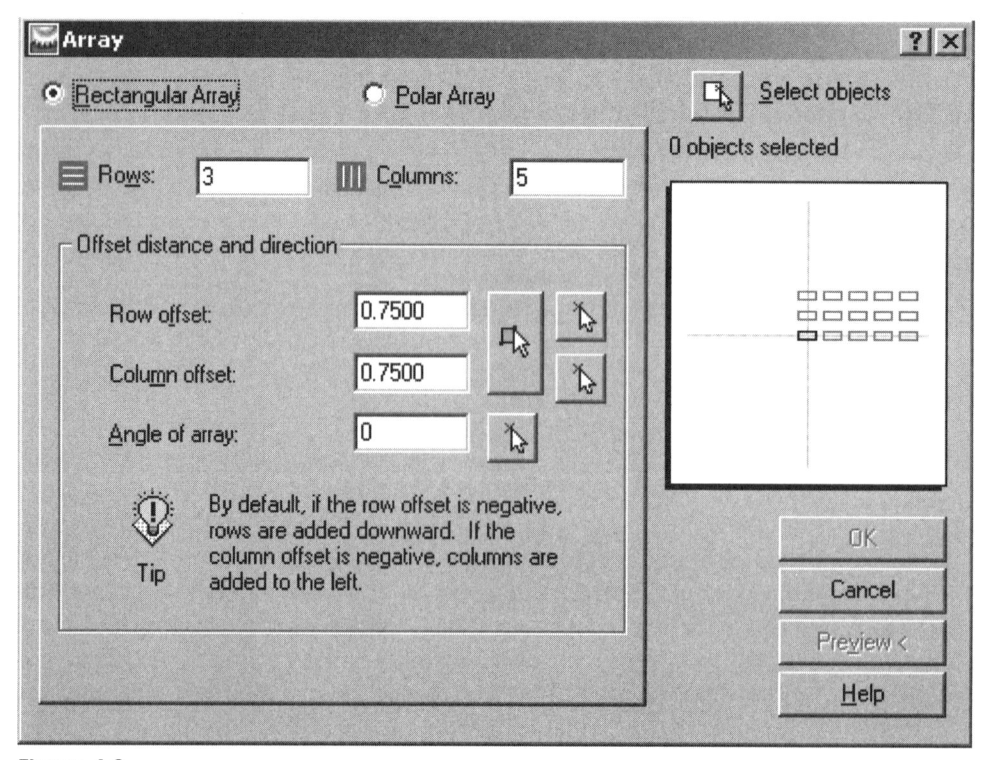

Figure 6.9

A polar array requires you to specify a center point, the total angle to be filled, and the number of items (or angle between items). The dialog box looks like Figure 6.10.

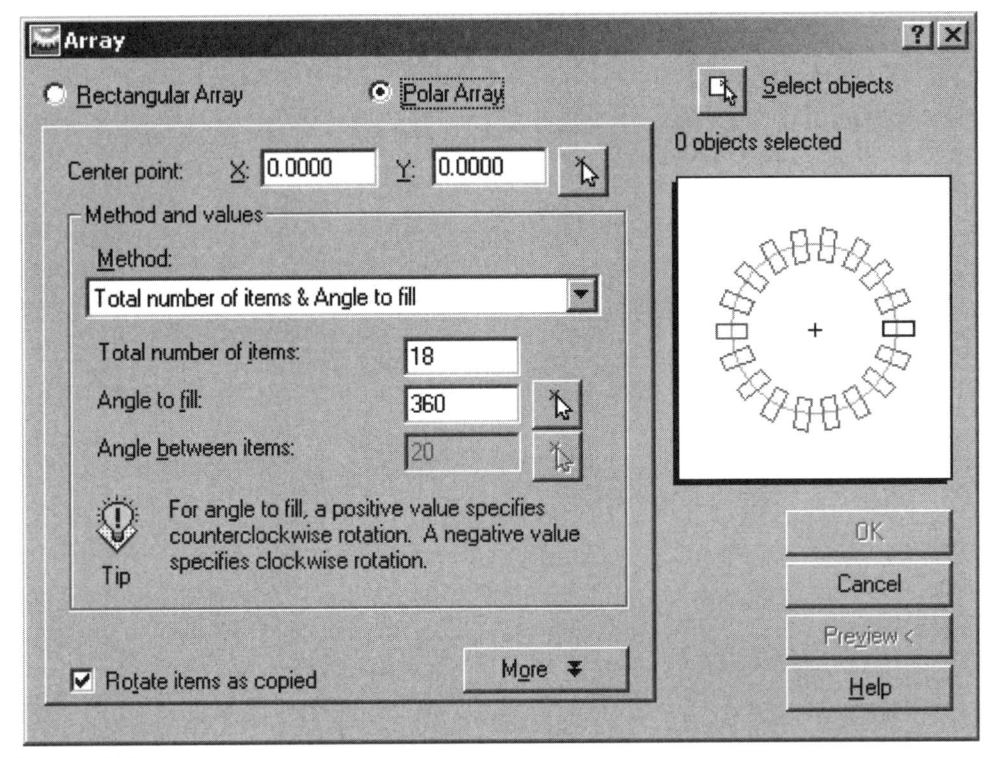

Figure 6.10

And, if completed as shown above, with the two lines selected, will convert the figure on the left into the one on the right as shown in Figure 6.11.

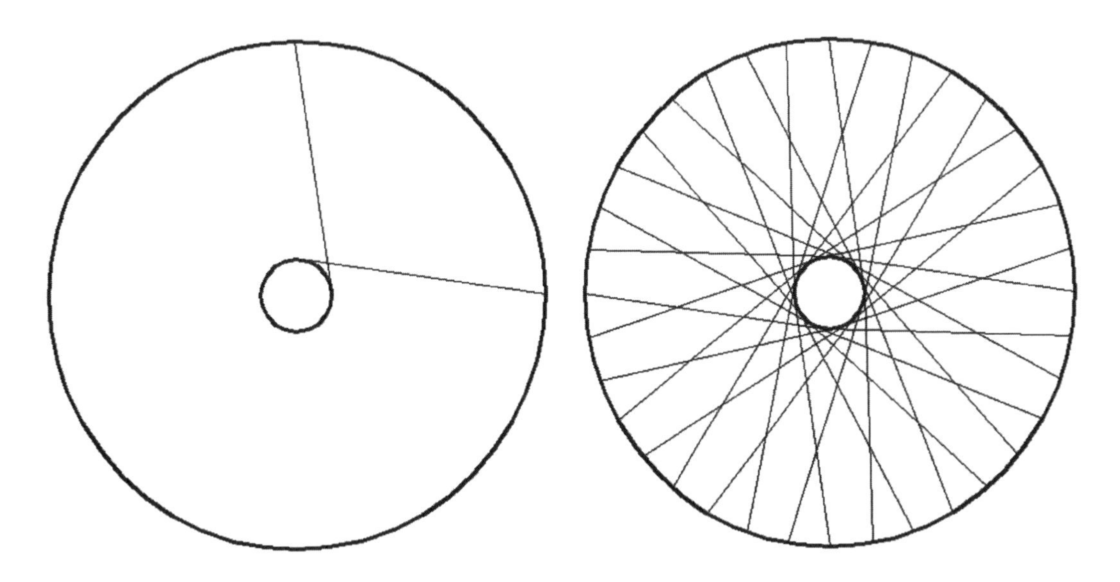

Figure 6.11

One particularly useful application of the array command is to draw threads on a bolt or inside a nut. This application combines keyboard coordinates and array to generate detailed threads on a bolt.

Problem: Draw detailed threads for the 1.5-6UNC-2A bolt shown in Figure 6.12. The length of engagement is 3.25".

Figure 6.12

The process we will follow is to draw one portion of the thread and then array it with one row and many columns to create the thread symbol. To begin the drawing we start a new drawing and draw one copy of the first thread.

```
Command: LINE
Specify first point:
Specify next point or [Undo]: @1/6<60
Specify next point or [Undo]: @1/6<-60
Specify next point or [Close/Undo]: @-1/12,1.5
Specify next point or [Close/Undo]: @1/6<240
Specify next point or [Close/Undo]: @1/6<0
Specify next point or [Close/Undo]: @1/6<120
Specify next point or [Close/Undo]:
```

Before going any further, it is important to understand what the keyboard coordinates used above actually did, and why. The first two sets drew the lower "V". Since we know the thread has 6 threads per inch (from the 6UNC portion of the note) we know the pitch is 1/6 of an inch and a UNC thread has a 60° thread angle. This means that the sides of the "V" make an equilateral triangle, with each side being 1/6". The next entry "@1/12,1.5" is probably the most confusing one. It draws the crest line on the thread, going backward in the "X" direction one half of a thread (1/12, not 1/6) and up the major diameter of the thread (from the 1.5-portion of the thread note). It is a relative Cartesian format. The next three entries draw the triangle at the top.

Having completed this your drawings should look like Figure 6.13. There is one extra line, the horizontal one, which we drew for construction. It should now be erased.

Figure 6.13

```
Command: erase
Select objects: {select short horizontal line} 1 found
Select objects:
```

We now have to add the root line, which must be added using OSNAP to find the exact root points on the "V"'s of the thread.

```
Command: LINE
Specify first point: endp of
Specify next point or [Undo]: endp of
Specify next point or [Undo]:
```

Figure 6.14

This should result in Figure 6.14, which is one copy of the thread. It now needs to be arrayed along the length of the bolt. You must first figure out how many copies of the objects you will need. Since the length of engagement is 3.5", and each thread is 1/6" long, you will need 3.5*6 copies, or 21 copies of the objects as columns and one row. With a column spacing of 1/6". Use the array command and complete the dialog box as shown in Figure 6.15.

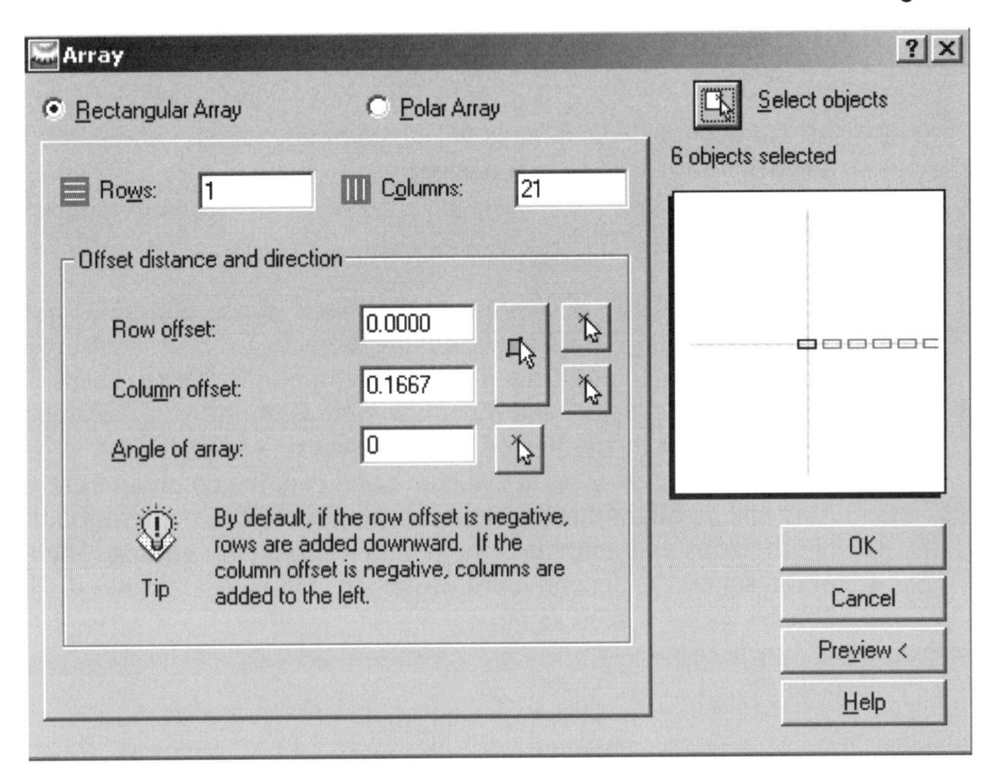

Figure 6.15

You should get the following results (Figure 6.16):

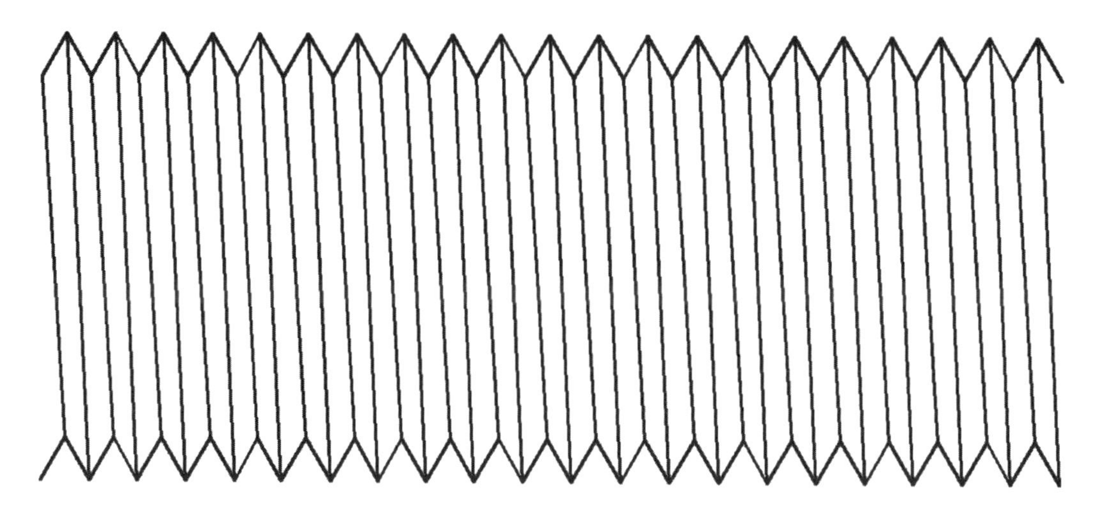

Figure 6.16

All that has to be done now is to complete the end of the bolt with a chamfer, then add whatever head type is desired.

Grips

In addition to the commands discussed above, AutoCAD has another method to allow you to make changes on the objects in your drawing. Grips can be used to accomplish some of the more common editing tasks. These include: Move, Copy, Stretch, Rotate, Scale, and Mirror as well as giving you the ability to move an object from one layer to another. To activate the grips on an object, you must left click on an object while there is no active command (i.e. the command prompt is "Command:"). This will highlight the object and display small blue squares on it. These blue squares are the grips and they mark the points available for editing. The number and location of the squares is dependant of the object. Figure 6.17 shows the grip locations for the objects we have seen so far.

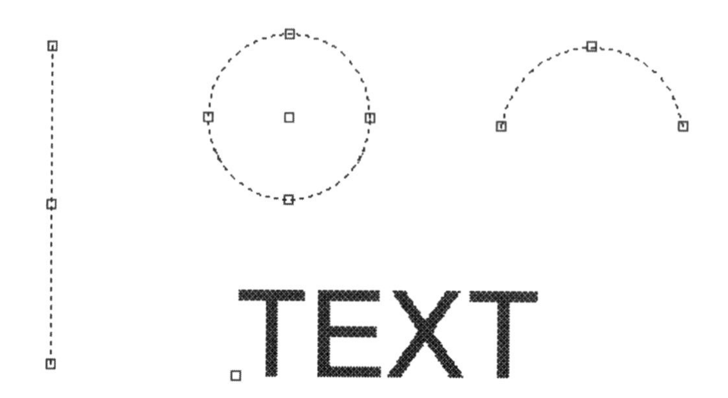

Figure 6.17

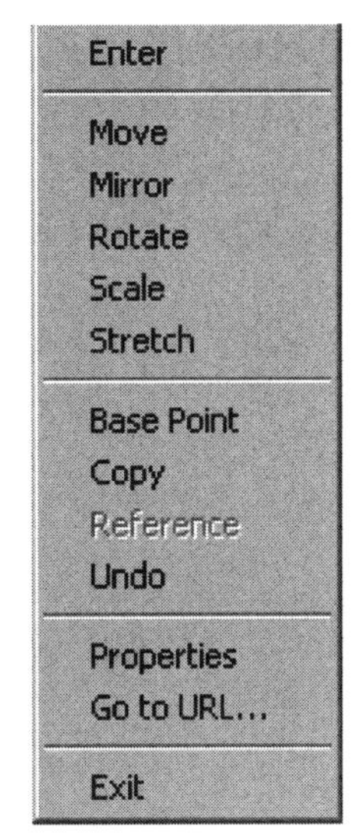

Figure 6.18

The blue grips are called "cold grips" and are just markers. To activate a grip and make it a "hot grip" click on that grip and it will turn red and be filled in and the command line will offer you the chance to stretch that grip to a new location. By pressing the enter key you can cycle through the five editing options, or you can right click and select them directly from the popup menu shown in Figure 6.18.

Once you activate grips on an object, the layer box will show whatever layer that object is drawn on (if you activate more than one object and they are on different layers, the box will go blank). To change that object to a new layer, just drop down the list of layers and select the new layer.

When you have finished making your modifications with grips, you can clear them by pressing escape on the keyboard two times (the first time will unhighlight the object without clearing the grips). If you invoke an editing command while the objects are still highlighted with grips, then the highlighted objects will become the selection set used for that editing command and no further selection prompt will be made. This can be very helpful at times, but you can get into trouble if you inadvertently forget to cancel grips and then use the ERASE command. If this happens, don't panic, just remember the Undo command and you can recover from your mistake.

Editing of Text

Text can be edited by using the command DDEDIT. This will request that you select a text string. The contents of that selection will be displayed in an edit text dialog box (Figure 6.19) and you can make whatever changes necessary. This is a common method when filling out title blocks. Rather than recreating each text entry from scratch, simply edit the existing text to add your name.

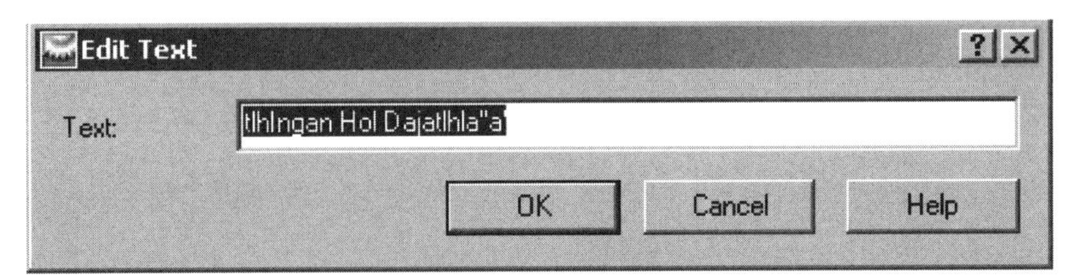

Figure 6.19

One other text operation available in AutoCAD is basic spell checking. The SPELL command will check a selection for spelling errors and display the proposed changes.

Object Properties

Each object created has certain properties associated with it. Many of these can be modified directly by using the command PROPERTIES. This command will activate a dialog box of sorts. It can be moved and docked at the side of the screen and will contain properties for editing, depending on what object(s) are selected using grips. Figure 6.20 shows 4 different object types: Line, Circle, Arc, and Text and the properties available for each.

Figure 6.20

Crosshatching for Detail and Effect

7

Objectives

1. Be able to apply crosshatching to objects in AutoCAD
2. Be able to correctly scale the crosshatching
3. Be able to edit the properties of crosshatching after it is applied
4. Be able to draw rectangles, ellipses, and polygons

For anyone who has done extensive hand drafting, one of the most onerous tasks is crosshatching a large area. No matter how hard you try, the lines never come out just right and the pattern seems to shimmer due to the inconsistencies within it. However, the process of applying a hatch pattern to an area is very methodical. This is another area, like editing, where the power of a CAD program will become obvious. One thing computers do very well is to follow an algorithm and applying a pattern is very methodical.

Defining the Boundary

In order to fill an area with a pattern, you much first define what area you want to fill. This is done by drawing objects to enclose the desired area. You must have a closed area defined, however the objects can overlap without causing a problem. Figure 7.1 used a line and a circle to crosshatch a portion of the circle.

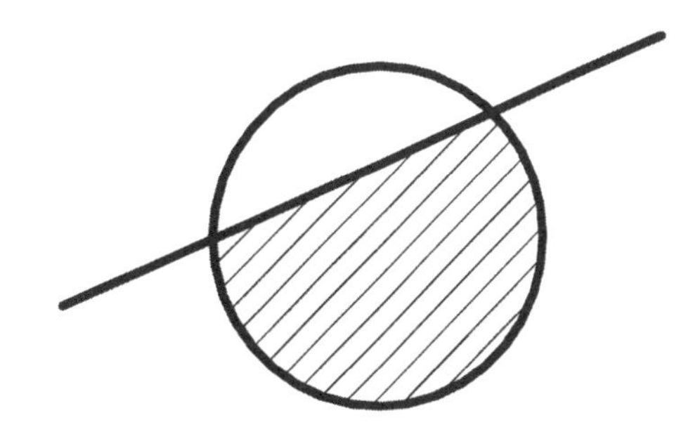

Figure 7.1

Notice that the line extends beyond the edge of the circle so that the area to be hatched is confined by part of the line and part of the circle.

Applying the Crosshatching

The command to actually apply crosshatching is BHATCH, which is located on the drawing toolbar. It will launch the dialog box shown in Figure 7.2.

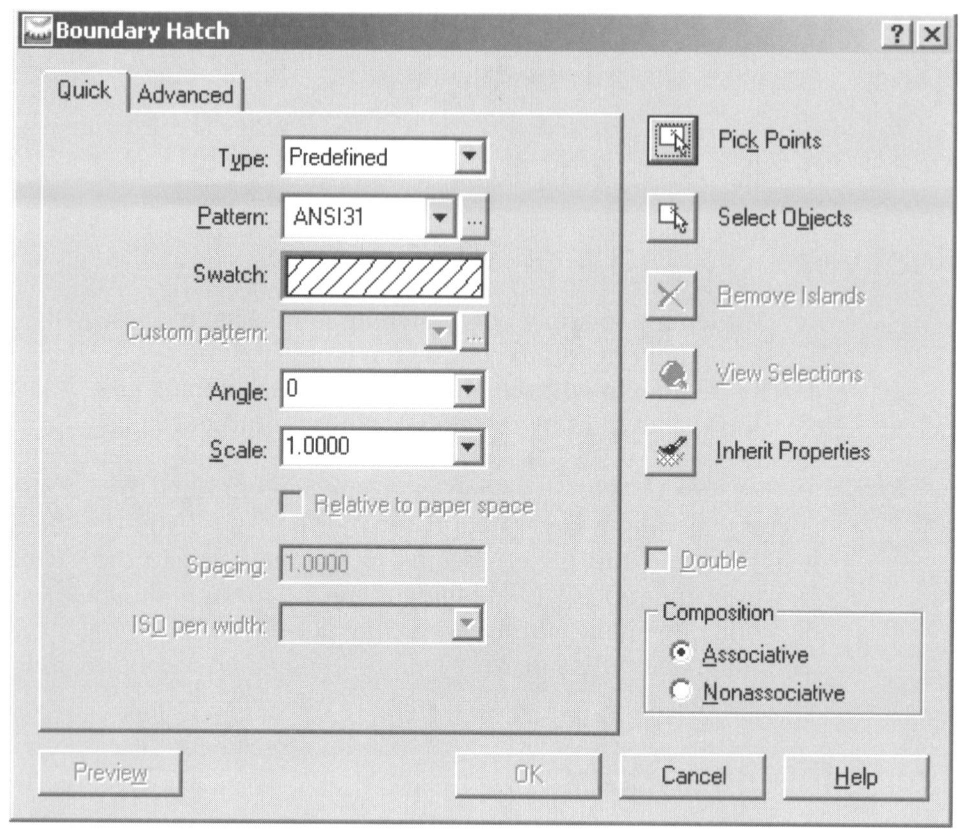

Figure 7.2

First, you should select what pattern you need to use. AutoCAD comes with 69 different patterns defined, and more can be downloaded via the Internet. The danger with that is many people go hog wild and start to pick all kinds of fancy patterns to "make the drawing look cool." An engineering drawing is a specific type designed to convey information. When you select a pattern, it should be representative of the material the object is constructed from. The most common materials in engineering and their pattern names in AutoCAD are shown in Figure 7.3.

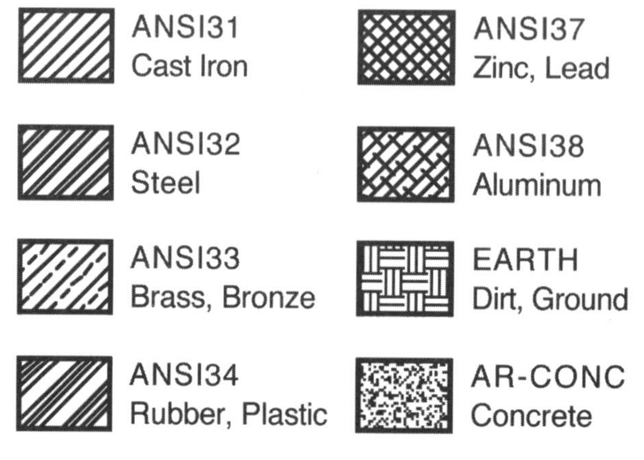

Figure 7.3

If you want to look through the available patterns then instead of just dropping down the list of patterns, select the ellipsis (. . .) to the right of the pattern name and you will be shown thumbnails of the available patterns.

With the pattern selected, you need to decide on the relative pattern scale and the rotation angle that should be used. The ANSI patterns are defined as full scale patterns and therefore, if your drawing is intended to be plotted at full scale you should leave the scale at 1 (or close to it). If the drawing is to be plotted to another scale, then the scale factor should be adjusted to account for plot scale. Non-ANSI patterns are drawn at irregular scales and may or may not look good at a scale factor of 1 on a full sized drawing. Especially the patterns prefixed with “AR-” the scale should be adjusted to a rather small value (about 0.03) for full scale drawings. The patterns are shown rotated with a zero degree rotations, you should change the rotation angle to account for different parts or irregular boundaries if necessary.

Selecting the Pattern, Scale, and Angle will determine the appearance of the crosshatching, now all that remains is to select where the pattern should appear. The most common manner to select this is to use the Pick Points option and select a point internal to the area you wish to crosshatch. AutoCAD will trace a boundary comprised of objects (or portions thereof) that enclose the point you selected. It will also select all closed boundaries within that first boundary.

These interior boundaries are called “islands”. Depending on the settings under the advanced tab they may or may not be crosshatched. Under the advanced tab shown in Figure 7.4, you may select whether AutoCAD should ignore these islands, only crosshatch the outermost ring available, or alternate hatching and not hatching the rings.

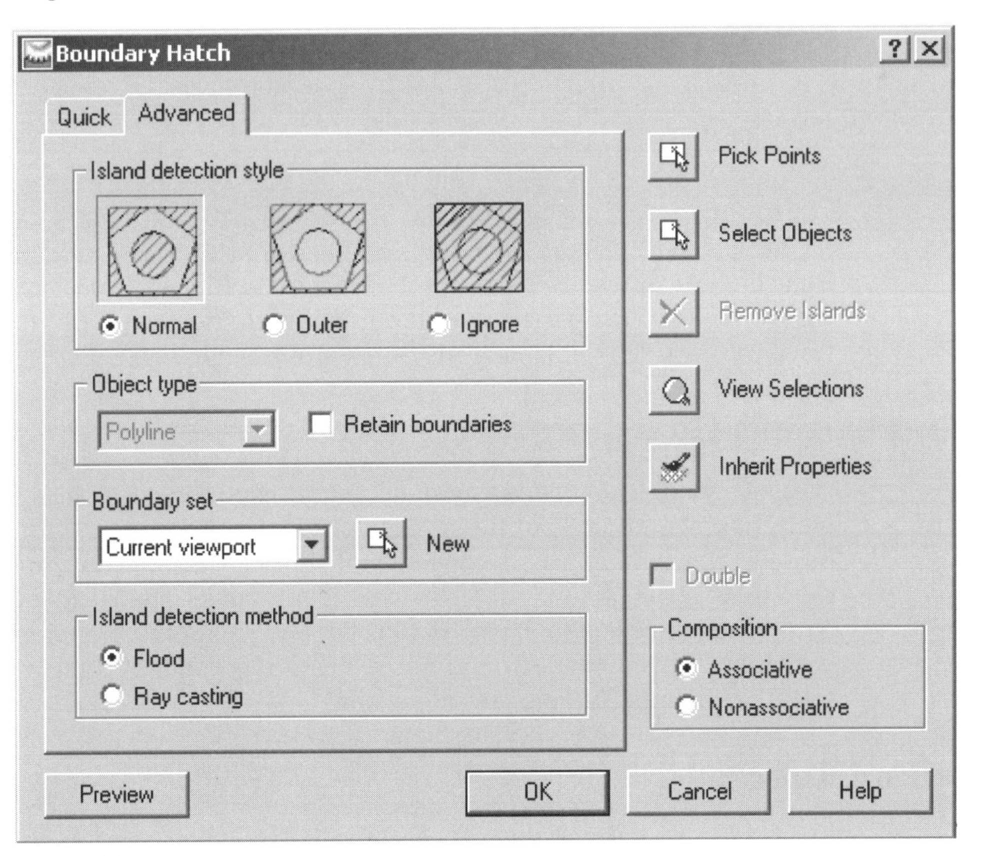

Figure 7.4

Once you have the boundary selected and the Island Style selected, preview the resulting pattern by selecting the Preview button. If you want to make any changes, press enter or right click to return to the dialog box and edit the parameters as you need. Once the preview looks the way you want it to, select OK and the crosshatching will be applied to your drawing.

If you already have some crosshatching on your drawing which you would like to match, you may select the Inherit Properties button and select the existing hatching. The Pattern Name, Scale, and Angle will be filled in to match the existing hatch.

Associative Crosshatching

One nice feature of AutoCAD's crosshatching is the associative nature of the pattern. The crosshatching is truly associated with the objects which form its boundary. If you edit one of these boundary objects, AutoCAD will try to update the crosshatching to accommodate the new position. If it cannot (for instance, you erased one of the boundaries), then you will receive a warning that the associativity of the crosshatching was removed. This will sometimes yield unusual results. Reusing the initial circle and line figure, if the line is moved (via grips) to different positions you could get the following two figures shown in Figure 7.5.

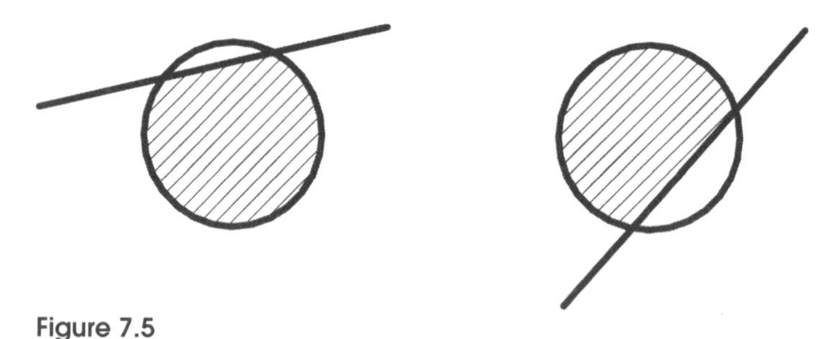

Figure 7.5

The figure on the left is what we would have expected, with the line moving farther up the circle the hatching expands to fill the new area, but the figure on the right is a bit different. That crosshatching has flipped from one side of the line to the other. The Line passed over the initial definition point of the crosshatching and thus, the area defined is now the portion of the circle shown.

Editing Existing Crosshatching

Once you have applied crosshatching, it is possible to make modifications in the pattern, scale, or angle of the hatching. The actual command to use is HATCHEDIT, however it is normally more convenient to double click the hatching which will launch the HATCHEDIT dialog box. This dialog box looks almost identical to the original apply hatch box shown earlier in the chapter. The only differences are in the buttons which are grayed out. You can not relocate the hatching, thus the "Select Objects" and "Pick Points" buttons are not available.

Special Objects

AutoCAD supports several special objects, which similar to circles, act as one object when being edited or crosshatched. These can be drawn with a single command and have multiple grips available for editing once created.

Rectangles

Rectangles are drawn using the RECTANG command or the icon from the draw toolbar. You will be prompted for opposite corners of the rectangle and it will be created with horizontal and vertical sides.

```
Command: rectang
Specify first corner point or
      [Chamfer/Elevation/Fillet/Thickness/Width]:
Specify other corner point or [Dimensions]:
```

Ellipses

Ellipses are slightly more complicated than rectangles to draw. You use the ELLIPSE command or the icon. There are several options for specifying the major and minor diameters of the ellipse. By default you will select the ends of one of the axes and then one end of the second axis. Depending on the relative size, the first may be the major or the minor axis. Optionally, you can specify the center of the ellipse first, then specify one end point of each axis. It is also possible to draw an elliptical arc using the ELLIPSE command.

```
Command: ELLIPSE
Specify axis endpoint of ellipse or [Arc/Center]:
Specify other endpoint of axis:
Specify distance to other axis or [Rotation]:
Command: ELLIPSE
Specify axis endpoint of ellipse or [Arc/Center]: C
Specify center of ellipse:
Specify endpoint of axis:
Specify distance to other axis or [Rotation]:
Command: ELLIPSE
Specify axis endpoint of ellipse or [Arc/Center]: a
Specify axis endpoint of elliptical arc or [Center]:
Specify other endpoint of axis:
Specify distance to other axis or [Rotation]:
Specify start angle or [Parameter]:
Specify end angle or [Parameter/Included angle]:
```

Polygons

Equilateral polygons are frequent objects needed on engineering drawings. Squares appear in many designs and hexagons are very common on bolts and nuts. To draw a polygon you use the POLYGON command or the icon. You will be asked how many sides you want your polygon to have. You may choose any integer between 3 and 1024, inclusive. You must then locate the polygon on the drawing by either specifying the location of the endpoints of one edge or the location of the characteristic circle associated with the polygon and whether the polygon is to be inscribed in the circle or circumscribed about the circle as shown in Figure 7.6.

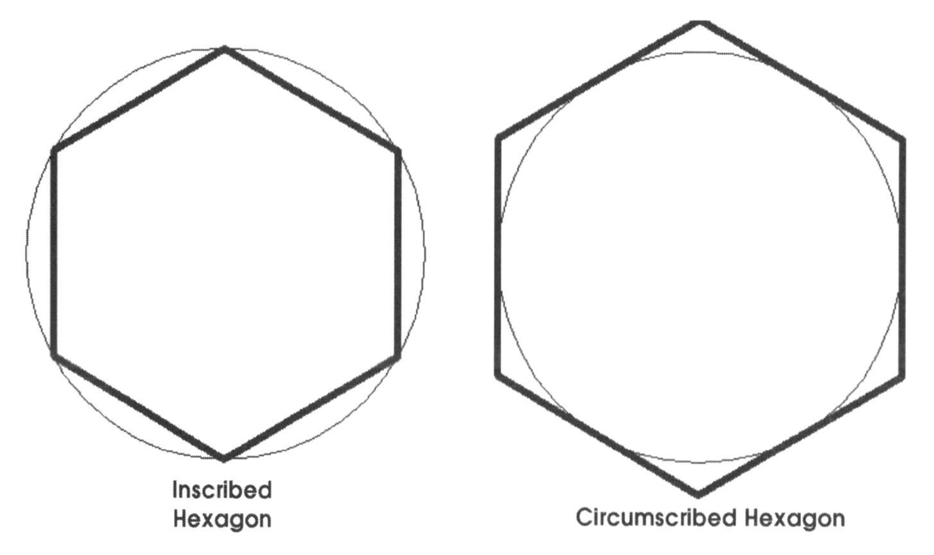

Figure 7.6

```
Command: POLYGON
Enter number of sides <6>: 8
Specify center of polygon or [Edge]: E
Specify first endpoint of edge:
Specify second endpoint of edge:
Command: POLYGON
Enter number of sides <8>: 6
Specify center of polygon or [Edge]:
Enter an option [Inscribed in circle/Circumscribed about circle]
      <I>: I
Specify radius of circle:
Command: POLYGON
Enter number of sides <6>:
Specify center of polygon or [Edge]:
Enter an option [Inscribed in circle/Circumscribed about circle]
      <I>: C
Specify radius of circle:
```

Dimensioning and Annotation of Drawings

Objectives

1. To be able to apply dimensions to drawings
2. To be able to apply centerlines to a drawing using dimensioning
3. To be able to apply linear tolerances to drawings
4. To be able to apply geometric tolerances to drawings
5. To adjust the dimensioning style to account for different plotting scales

One of the major purposes of an engineering drawing is to document the design for construction. In order to actually build an object, it is necessary to know how big it will be. Thus, engineering drawings require dimensions. Hand in hand with dimensions are tolerances. To function correctly, tolerances are frequently necessary.

AutoCAD does semi-automatic dimensioning. It will apply dimensions in any manner the user tells it, without regard to proper style, however all the measurements will be correct. As can be seen from the toolbar shown in Figure 8.1, there are many options for applying dimensions in AutoCAD. In each case, you should have some idea about what your finished drawing should look like prior to applying dimensioning.

One point to make about the method AutoCAD uses to apply and calculate dimensions for a drawing. By default, AutoCAD creates dimensions that dynamically update if you edit the size or location of a feature on the drawing. This is controlled by a parameter in AutoCAD called DIMASSOC. This parameter has 3 valid settings: 0, 1, and 2. The setting of 0 will disable all the updating capability of the dimensions and remains in AutoCAD only to maintain backward compatibility. Use of mode 0 is not recommended. Mode 1 allows fully associative dimensions in Model Space. Mode 2 is new to AutoCAD 2002 and supports advanced features like dimensioning in Layout mode. This chapter assumes that you have set DIMASSOC to 2. If you are following along using AutoCAD and are getting different results than presented in the examples, check this setting.

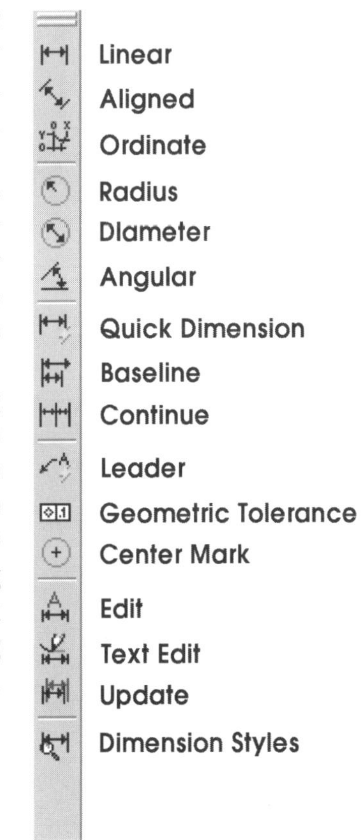

Figure 8.1

Linear Dimensions

Linear dimensions comprise most of the dimensions for engineering drawings. They are defined by two points for the ends of the extension lines, one point for the locations of the dimension line, and an angle for the dimension line. (Remember to keep terminology correct, what is a dimension line, and what are extension lines.) Normally the dimension line angle will be either 0° (horizontal) or 90° (vertical) and this orientation can be inferred by the selection of the three points.

As an example, look at applying dimensions to the sloped line in Figure 8.2. In each case, the 1st and 2nd points were chosen the same, but the 3rd point selected determined whether the dimension was to be a horizontal one (the 3.5) or a vertical one (the 1.5).

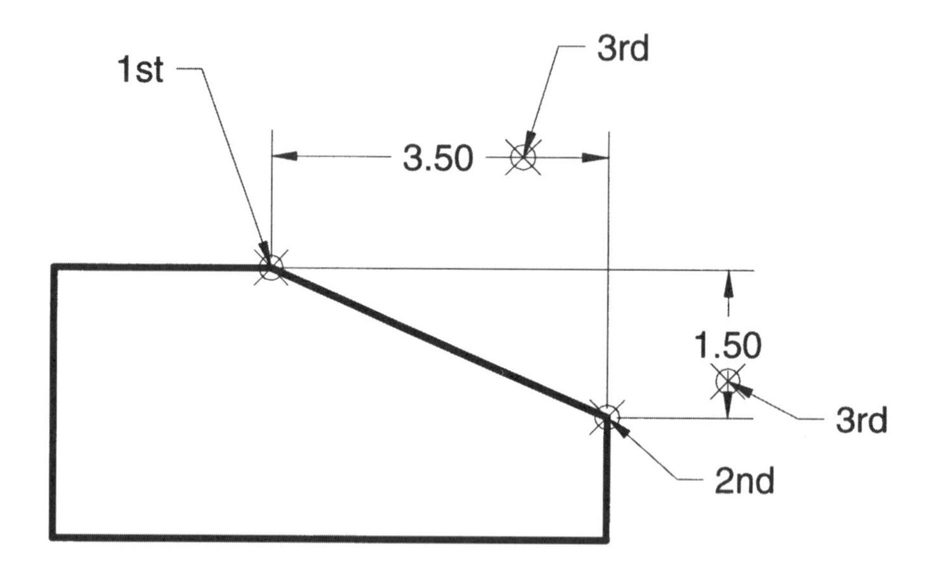

Figure 8.2

```
Command: _dimlinear
Specify first extension line origin or <select object>: {select
      point 1}
Specify second extension line origin: {select point 2}
Non-associative dimension created.
Specify dimension line location or [Mtext/Text/Angle/Horizontal/
      Vertical/Rotated]: {select point 3 on the 3.50 dimension}
Dimension text = 3.50
Command: _dimlinear
Specify first extension line origin or <select object>:{select
      point 1}
Specify second extension line origin: {select point 2}
Non-associative dimension created.
Specify dimension line location or [Mtext/Text/Angle/Horizontal/
      Vertical/Rotated]: {select point 3 on the 1.50 dimension}
Dimension text = 1.50
```

Notice that no options were selected to change from a horizontal to a vertical dimension. Prior to selecting the 3rd point, you may choose one of the available options, but in most cases, it is not necessary.

Aligned Dimensions

If you had wanted to dimension the actual length of the sloped face, then you would use an aligned dimension rather than a linear one. The result would be:

```
Command: dimaligned
Specify first extension line origin or <select object>:{select
      point 1}
Specify second extension line origin: {select point 2}
Non-associative dimension created.
Specify dimension line location or [Mtext/Text/Angle]: {select
      point 3}
Dimension text = 3.81
```

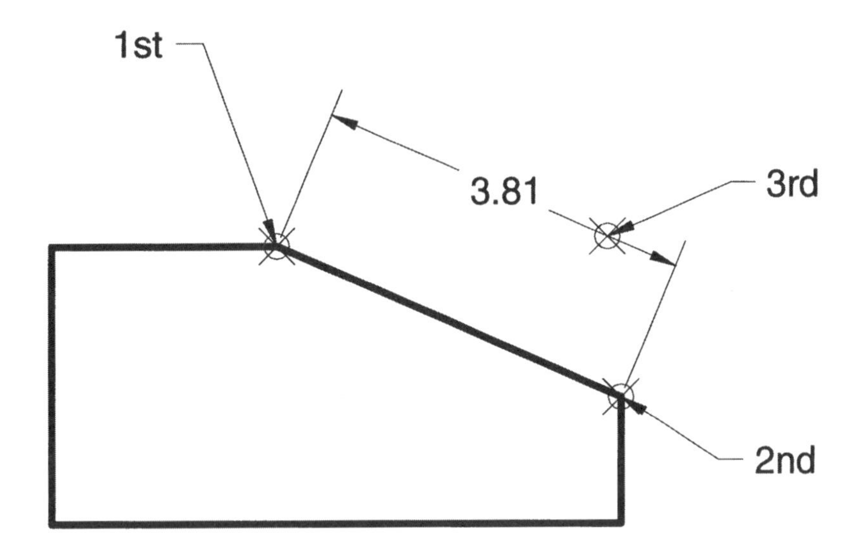

Figure 8.3

Radius and Diameter Dimensions

If you have an arc or a circle, then you should use either Radius (if it is an arc) or Diameter (if it is a full circle). The application of these dimensions is straight forward, but they open a virtual Pandora's box of options due to the presence or absence of centerlines. AutoCAD can be set to apply centerlines automatically when applying radius or diameter dimensions. An arc can be dimensioned with any of the following centerline combinations:

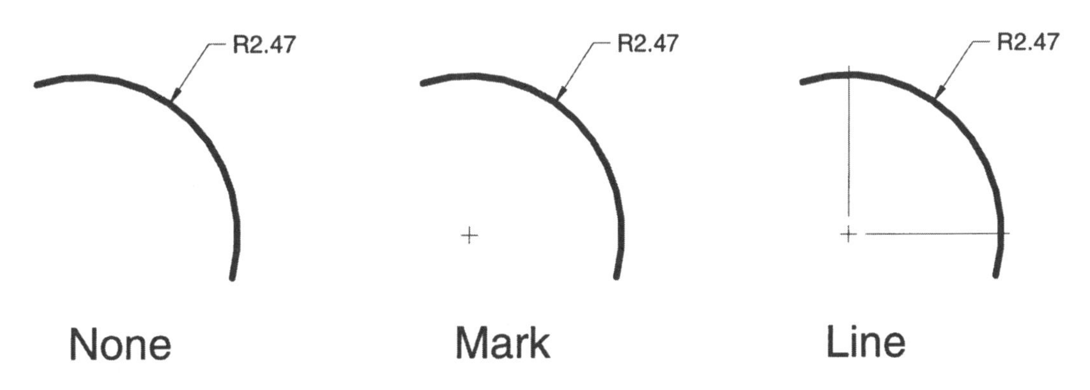

Figure 8.4

It is possible to have AutoCAD simply draw the centerlines without applying either a radius or diameter dimension by using the Center option and selecting the arc or circle.

These are controlled through something referred to as Dimension Styles. Like Text Styles they are a set of parameters used to define the appearance of dimensions. There are 70 different settings which affect the appearance of dimensions. These are all controlled through a series of dialog boxes accessed via the DIMSTYLE command. Most of the settings will be established in the base template file and you will not have to modify them, however a few are worth noting in detail. The initial dialog box is shown in Figure 8.5.

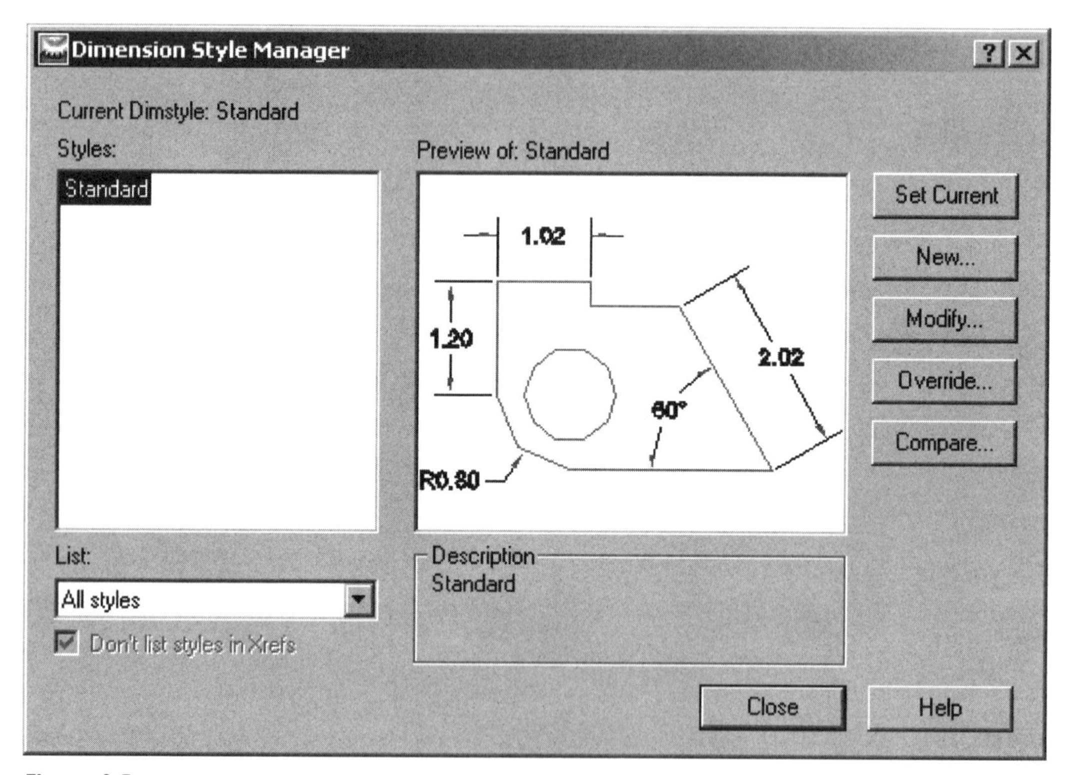

Figure 8.5

All the settings can be accessed by choosing modify. This will open a new dialog box with 6 tabs. The settings that you may want to adjust are Center marks (under the Lines and Arrows tab), Text Alignment (under the Text tab), Scale for dimensioning features and Fine Tuning (under the Fit tab), and Precision (under the Primary Units tab).

Center Marks: This setting controls how the Radius, Diameter and Center options draw center marks. If used with the Radius or Diameter options the center marks become part of the dimension, however if used with the Center option they are independent objects and can be modified as such.

Text Alignment: Probably does not need to be modified, since the engineering standard of Unidirectional is the default.

Scale for Dimensioning Features: This is where you would account for a plotting scale other than 1 to 1.

Fine Tuning: Normally AutoCAD will attempt to place the dimension directly in the middle of the dimension line. If you would rather have the text placed at the location of the 3rd selection point, then choose “Place text manually when dimensioning”.

Precision: This controls the number of decimal places AutoCAD uses for dimensions.

Angular Dimensioning

Angular dimensions can be applied to an arc, two points on a circle, two lines, or three points and will measure the angle associated with the objects.

Baseline and Continue Dimensioning

Baseline and continuation dimensioning are really just specialized forms of linear dimensioning. With a baseline dimension AutoCAD will hold the first extension line location in memory and repeatedly ask you for a second extension line location, drawing a dimension for the first to each successive second. Continue holds the second extension line in memory, transferring it to be a first and then prompts you for a new second extension line location.

Closely related to Continue dimensioning is Quick Dimensioning. Quick dimensioning prompts you to select the geometry you want to dimension. This can be done with any of the selection set techniques discussed for editing. Once you have everything selected that you want to dimension, press return and you can drag either a horizontal or vertical dimension in place for all of the objects concurrently. It is a neat option without a lot of practical applications.

Leaders

Leaders are a very flexible method for applying general notes to specific features on a drawing. They are very useful for thread notes, knurls, and chamfers. The leader begins on the feature to be dimensioned and ends at the location of the note.

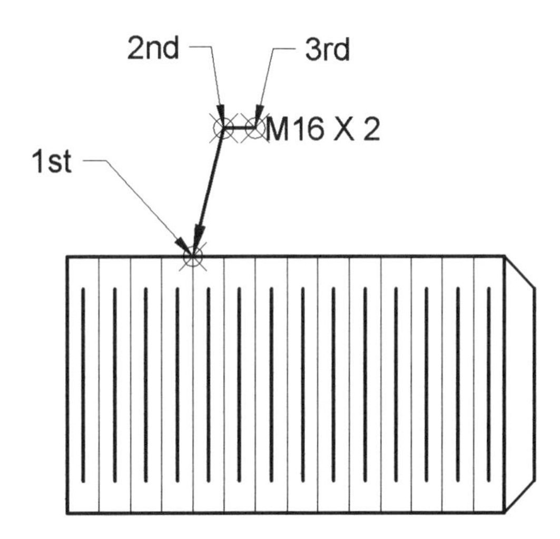

Figure 8.6

```
Command: _qleader
Specify first leader point, or [Settings] <Settings>:
Specify next point:
Specify next point:

Specify text width <0.0000>: {Press Enter}

Enter first line of annotation text <Mtext>: M16 X 2
Enter next line of annotation text:
```

Editing of Dimensions

Dimensioning styles are useful, but more often changes need to be made on one particular dimension, not globally on all dimensions. To do this you need to modify properties on the individual dimension. The same six categories are available for editing as were in the Dimension Styles, but these properties apply only to the dimension selected. If you need to change the text height for one particular dimension or the center mark style of one, you would modify the properties for that dimension using the Modify Properties command. The resulting properties box is shown in Figure 8.7.

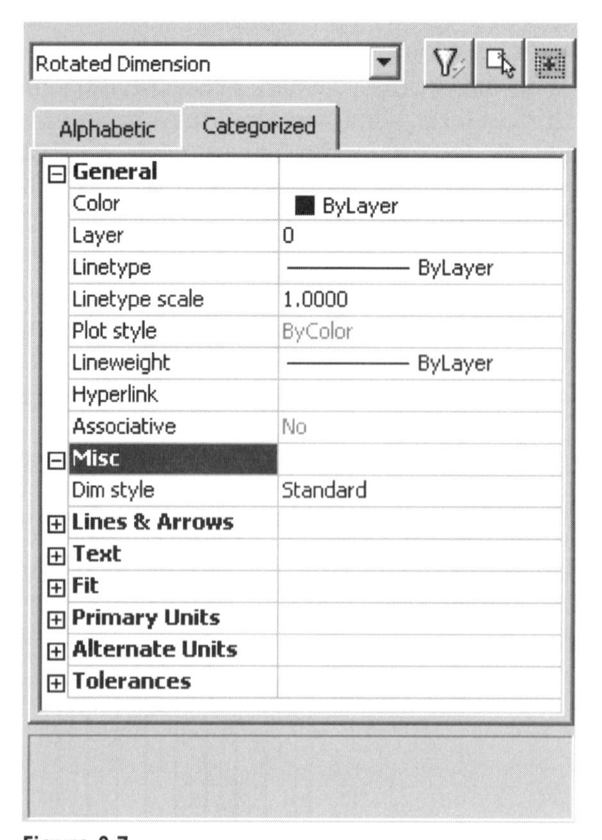

Figure 8.7

Linear Tolerances

From an engineering drawing point of view, one of the most useful applications of editing dimensions is to apply linear tolerances.

To apply the tolerance shown in Figure 8.8, you must first make sure the part is drawn **exactly** to the correct size. Do not try to draw a part without snap on and then apply dimensions to it, much less toleranced dimensions.

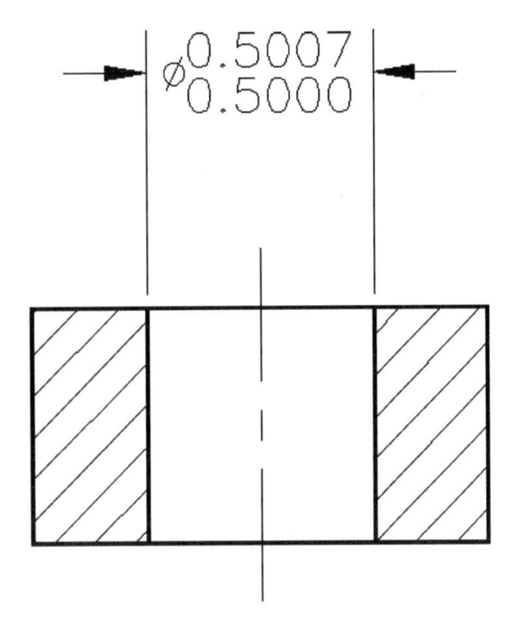

Figure 8.8

Apply a dimension of 0.5000 in the normal fashion. Then perform a modify properties operation on this dimension. Under the Tolerance sections are three parameters which need to be checked. The dialog box is shown in Figure 8.9.

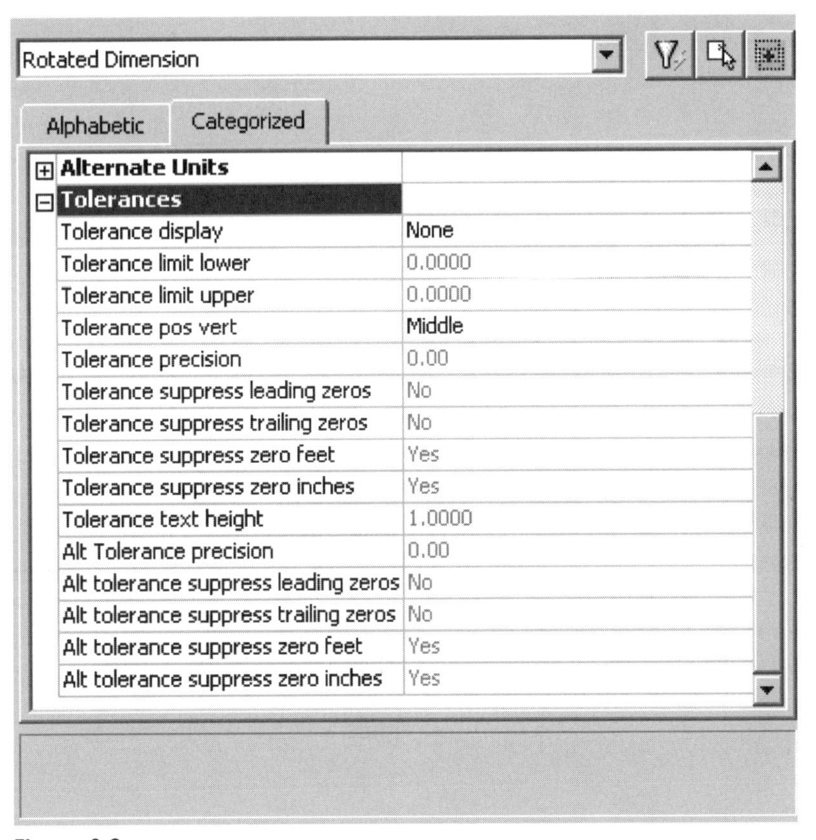

Figure 8.9

The tolerance display should be set to Limits form in order to display the tolerance correctly on the drawing. Setting it to Symmetrical or Deviation will tolerance the dimension, but it is not in the engineering standard format. Once that has been switched to Limits, then the "Tolerance limit lower" and "Tolerance limit upper" will ungray and you should enter the deviations from the basic size here. The "Tolerance Limit Lower" will be subtracted from the basic dimension to obtain the lower limit and the "Tolerance Limit Upper" will be added to the basic dimension to get the upper limit. For this dimension, the Tolerance Limit Upper should be set to 0.0007, and the Tolerance Limit Lower should be set to 0.0000.

Geometric Tolerances

Linear tolerances are only half of the picture. Geometric tolerances specify not the size of a feature, but the shape and orientation. AutoCAD has a special mode for applying geometric tolerances. The dialog box (Figure 8.10) guides you through the process very well.

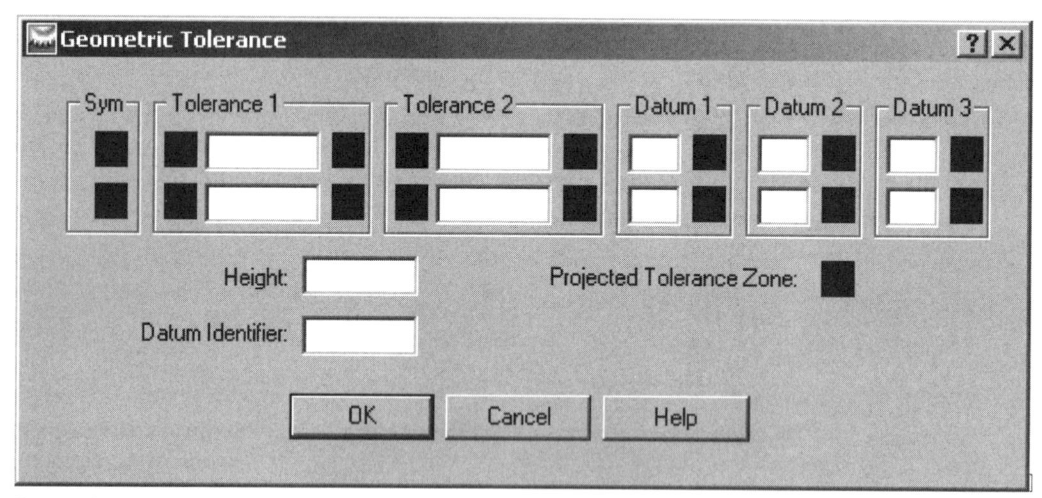

Figure 8.10

If you click on the "Sym" section, you will get another box of possible symbols shown in Figure 8.11. You just select the correct geometric tolerance symbol and it will be inserted at the correct location in the callout box.

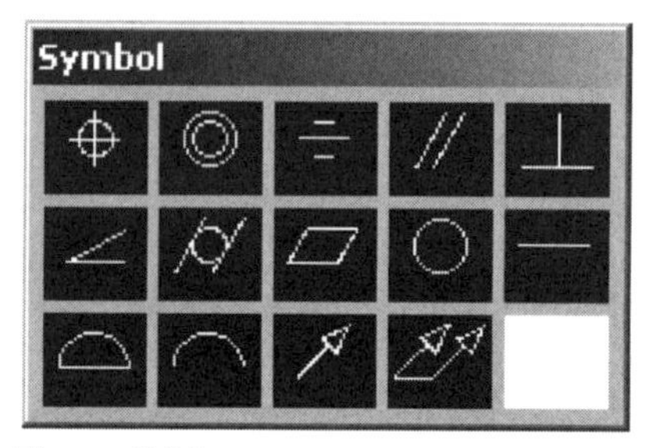

Figure 8.11

The square in front of the tolerance zone is for a diameter symbol "Ø" and by clicking on it you can toggle the display of that symbol. The box after Tolerance zone is for Material Condition specification (Figure 8.12).

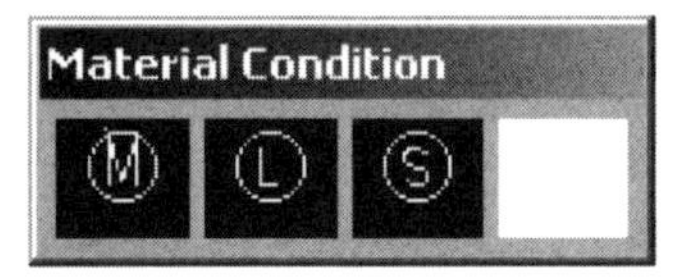

Figure 8.12

Once you get all the specifications entered into the box, select OK and you will be able to drag the box into place on the drawing. Notice that this does not draw a leader, it just draws the feature control frame.

Example:

Create the following annotations on a drawing.

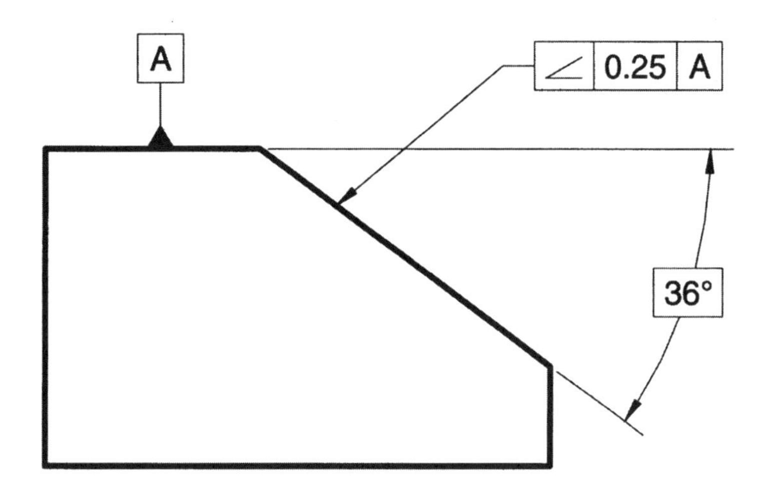

Figure 8.13

Step 1: Draw the figure

Step 2: Place an angular dimension between the top surface and the sloped surface. Do not worry about the basic dimension indicator for the moment.

Step 3: Add the basic dimension indicator to the angular dimension by selecting it and doing a modify properties. Under the Tolerance section select the "Tolerance Display" and choose "Basic." This will place a box around the dimension indicating that it is a basic dimension.

Step 4: Add the datum reference to the upper surface.

Step 4A: Turn ortho on and select the Quick Leader icon from the dimension toolbar (or use QLEADER). Before selecting any points go into the setting option. You will see a dialog box like the one shown in Figure 8.14.

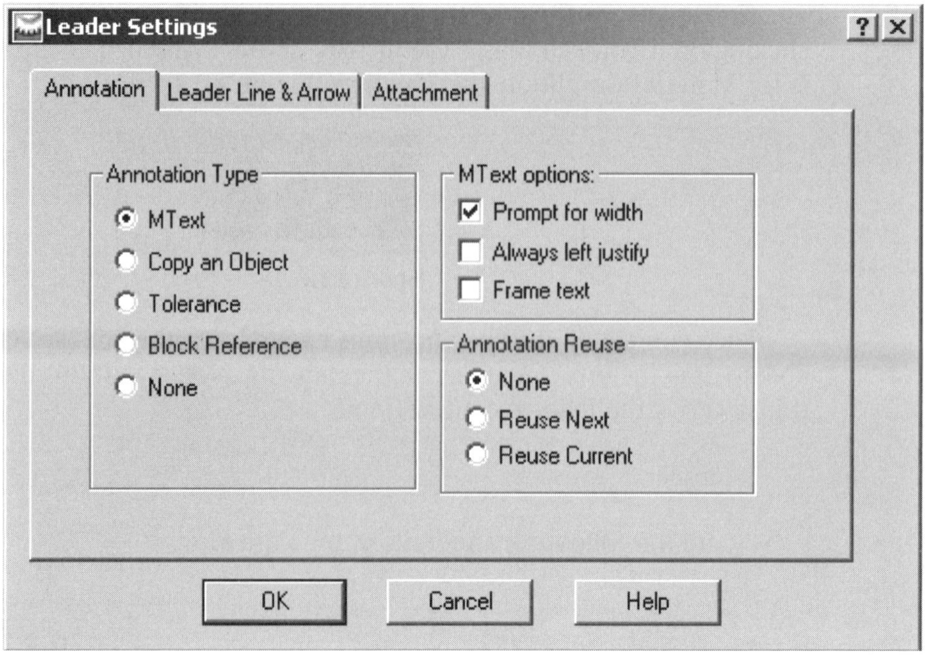

Leader Settings
Annotation
Leader Line & Arrow
Attachment
Annotation Type
MText
Copy an Object
Tolerance
Block Reference
None
MText options:
Prompt for width
Always left justify
Frame text
Annotation Reuse
None
Reuse Next
Reuse Current
OK
Cancel
Help

Figure 8.14

Set the Annotation type to None and on the "Leader Line & Arrow" tab select "Datum Triangle Filled" and select OK. Now pick a point on the upper surface and one directly above it. Then press return to complete the Datum Leader.

Step 4B: Select the Tolerance option from the Dimension Toolbar and enter just the letter "A" in the lower "Datum identifier" position. Select OK and drag the datum identifier into place. (You may have to get it close and then move it exactly into place using OSNAP midpoint on the box and endpoint on the leader.)

Step 5: Create the Feature Control Frame. Return to the Tolerance icon on the toolbar and enter the information for the Feature Control Frame as shown in Figure 8.15. Then select OK and drag it into place.

Geometric Tolerance
Sym
Tolerance 1
0.25
Tolerance 2
Datum 1
A
Datum 2
Datum 3
Height:
Projected Tolerance Zone:
Datum Identifier:
OK
Cancel
Help

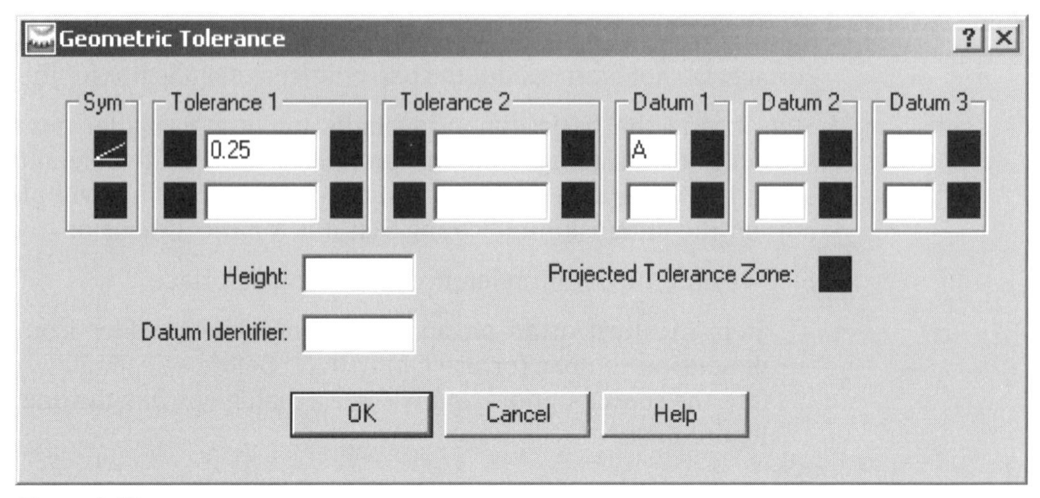

Figure 8.15

Step 6: Add the final leader. Select the Quick Leader option and return to the settings option. Reset the arrowhead to a normal Closed Filled arrow and apply the leader beginning on the sloped surface and ending at the Feature Control Frame.

Merging of Files, Blocks, and External References

Objectives

1. Be able to create a Block in AutoCAD
2. Be able to insert a Block, from disk or memory, into the current drawing
3. Be able to drag objects from one open drawing into another
4. Be able to place a bitmapped graphic image on an AutoCAD drawing
5. Be able to use files as external references

As stated earlier, one of the advantages of CAD is that you never have to draw anything twice. Once you have drawn something in one place, you can simply copy it to another location and avoid the time and effort of recreating it. As long as it is within the same file this works, but between files it requires a different approach. One of these different approaches is to use Blocks to copy items. A block is simply a collection of objects grouped together and given a name. This group effectively becomes a new object type which can be placed in the drawing as if it were any of the simple objects we have examined so far.

Creation of Blocks

The hardest part of block creation is simply drawing the objects you want to group together. These are drawn just as if you were trying to draw the symbol on the drawing normally, with no concept of making a block of it. This statement is not quite true, since there are a few things that you can, but do not have to, consider. These deal with putting the objects on the correct layer so that they will behave as you want in subsequent insertions of the block.

The command to create the block is BLOCK and it will launch the dialog box shown in Figure 9.1.

The first requirement is that you specify a unique name to identify the block. This name will be the reference for the block in the future. You must also select the objects you want to include in the block. This is the same as forming a selection set for editing. Finally, you must select a Base Point. To understand this we need to look toward the future a little and see how the block will be used.

If you have created a block of the end view for a hex nut, when you place the nut on a drawing you will do so by picking a point to locate the hex nut. The question is, "What point will that be on the Hex Nut?" Will it be the center of the hole? Will it be the center of one side? Will it be the left corner of the Nut? Any of these are possibilities, so when the block is made, you must specify the base point so that AutoCAD will know which point to reference when inserting the block.

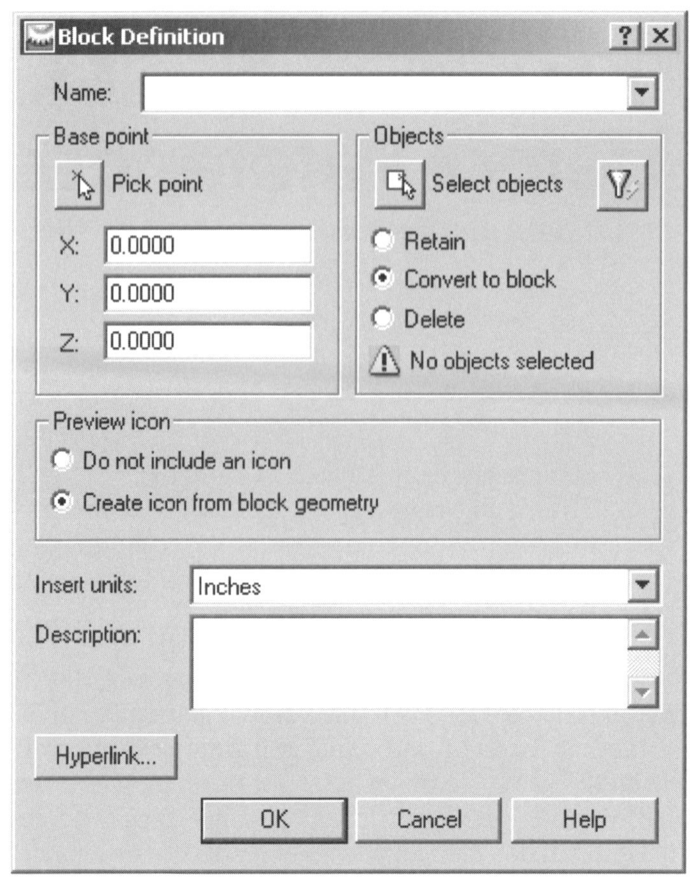

Figure 9.1

To actually select the base point you can either enter the coordinates directly in the spaces provided on the dialog box, or select the Pick Point button and actually select the point on your drawing.

Once these have been selected, selecting OK will create the block. Depending on the option selected under the Objects section, the original will either be deleted, retained, or converted to a block.

Inserting a Block

With a block defined we can now discuss placing the block elsewhere in the drawing. The command to place a block in a drawing is INSERT. As usual, you will see a dialog box appear (Figure 9.2). There are a few interesting features that are available when inserting blocks. They are not just exact carbon copies of the original definition but they can be scaled larger or smaller (independently in X, Y, and even Z) or they can be rotated from the original definition. This gives you a lot of flexibility since you don't want to have to draw every bolt you might need in all its orientations, you want to draw it once and then block and insert it as needed at whatever rotation angle each occurrence requires.

There is also a Browse... button. It is possible to insert an entire file into your current drawing. The whole file will become a single block in the current drawing and can be moved, scaled, or rotated as one unit. If you want to be able to operate on individual objects within that block, check the Explode box in the lower left corner. This will force the block to appear not as a single unit, but as a bunch of individual objects.

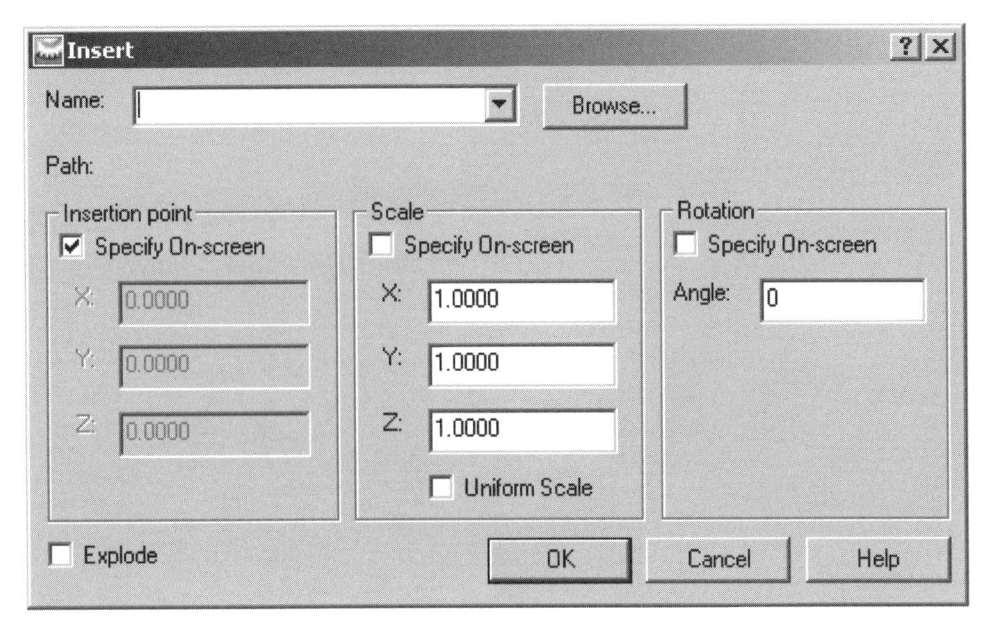

Figure 9.2

Example: Draw the end view of a 1-12UNF-2B hex nut (Figure 9.3).

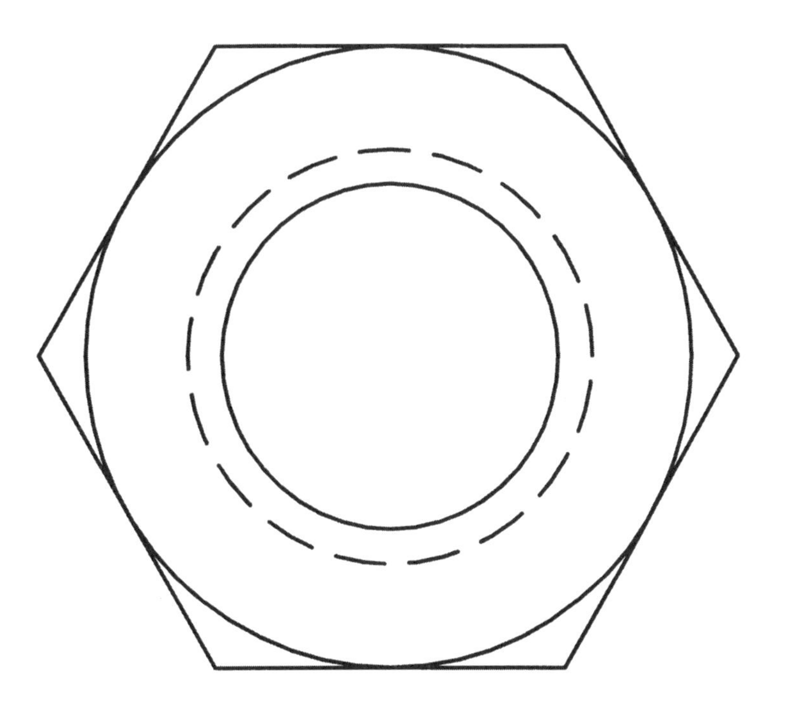

Figure 9.3

Set the snap and grid for the drawing. Also set the current layer to Visible.

```
Command: snap
Specify snap spacing or [ON/OFF/Aspect/Rotate/Style/Type]
      <0.5000>: 0.125
Command: grid
Specify grid spacing(X) or [ON/OFF/Snap/Aspect] <0.5000>: 0.25
```

Draw the circle for the major diameter.

```
Command: circle
Specify center point for circle or [3P/2P/Ttr (tan tan radius)]:
      1,1
Specify radius of circle or [Diameter]: 0.5
```

Create the minor diameter by offsetting the major diameter by the pitch.

```
Command: offset
Specify offset distance or [Through] <1.0000>: 1/12
Select object to offset or <exit>: {select the circle}
Specify point on side to offset: {select a point inside the
      circle}
Select object to offset or <exit>:
```

Change the major diameter circle from visible to layer hidden by selecting it using grips and changing the layer via the dropdown list of layers. This step does not give any command line feedback and is, therefore, not noticeable in the command stream.

Draw the outer circle of the nut. Remember that for a normal hex nut the outer circle has a diameter of 1.5 times the major diameter of the thread.

```
Command: circle
Specify center point for circle or [3P/2P/Ttr (tan tan radius)]:
      1,1

Specify radius of circle or [Diameter] <0.5000>: 0.75
```

Add the surrounding hexagon.

```
Command: polygon
Enter number of sides <4>: 6
Specify center of polygon or [Edge]: 1,1
Enter an option [Inscribed in circle/Circumscribed about circle]
      <I>: c
Specify radius of circle: .75
```

Make the block using the dialog box settings shown in Figure 9.4.

```
Command: block
Select objects: Specify opposite corner: 4 found
Select objects:
```

Drag and Drop

If you want to move a portion of one drawing into another, there are several strategies. You can use the command WBLOCK to write a portion of your current drawing to disk as a separate drawing. This will allow you to insert that disk file in another drawing or just allow you to split a drawing into smaller pieces. While this process works to transfer objects, if that is your only goal, there are easier methods.

If you open both the file which contains the objects you want to move and the file that you want to move them into, the objects can be selected with grips and then you can drag them from one drawing into another. Highlight all the objects you want to move and then select one of the grips using the right mouse button. Hold the right

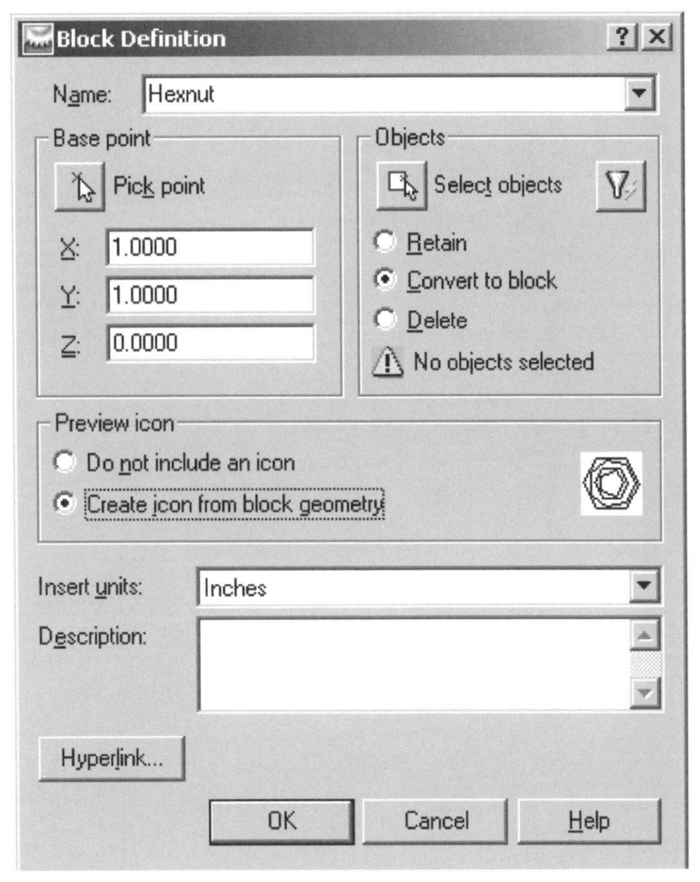

Figure 9.4

button down and you can drag the items to a new location within that file, or into the adjacent file. When you release the right button, you will be prompted with the popup menu shown in Figure 9.5. You can copy the objects or make a block of them in the new file.

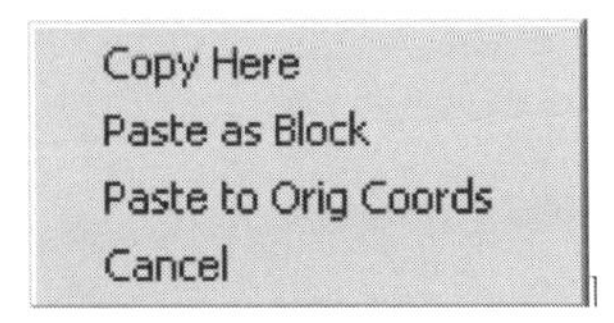

Figure 9.5

Windows Cut and Paste

You can also use the traditional Windows Cut and Paste to copy objects to the clipboard in one drawing and to paste them to the drawing area in another.

3D-Modeling—Regions and Booleans

Objectives

1. Be able to create regions from existing objects
2. Be able to create regions using the boundary command
3. Be able to extract the physical properties of a Region
4. Be able to move the origin of a drawing
5. Be able to list the physical properties which are dependent on origin location
6. Be able to annotate the drawing with that information
7. Be able to use the Boolean operations on existing regions

By this point, you know enough about AutoCAD to create a set of working drawings for most any item. An important point to remember about CAD is that the full acronym is CADD—Computer Aided Design and Drafting. The concept of design, that is using the computer to assist, not in the drawing for documentation, but in actually helping with the work of designing an invention. This is where the realm of 3D drawing becomes most useful and to introduce the concepts that underlie 3D AutoCAD, we should first investigate Regions.

Creating Regions

Regions are planar objects which enclose an area. They must be formed by drawing a simple closed figure using normal 2D drawing objects. The resulting figure must be a non-overlapping closed boundary. Figure 10.1 shows several examples. Those on the left are valid objects to create a region and the ones of the right are not.

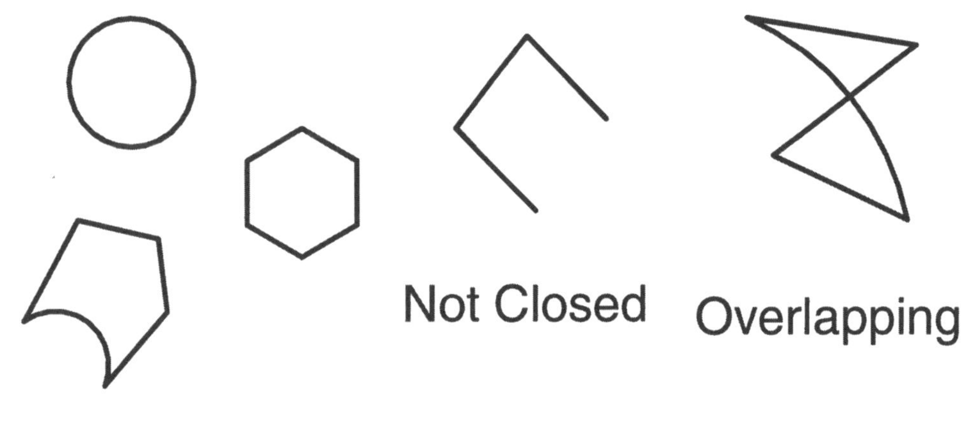

Figure 10.1

To create a region draw the figure using simple objects. Then convert those objects to a region using the REGION command. Unlike the BHATCH command, AutoCAD will not allow you to make a region from portions of objects using the REGION command, however, the boundary command will allow it. Just change the object type from Polyline to Region.

The following four attempts to create regions illustrate the possible outcomes of the REGION or BOUNDARY command. In the first, five objects were selected and they formed a valid region. Thus AutoCAD issued the message that it found one valid loop (boundary) and constructed one region from that loop. In the second, three objects which did not close were selected and AutoCAD could not find any valid loops. In the third, AutoCAD did find a closed loop, but the resulting region would have intersected itself, thus it was rejected. Finally, the boundary command was used and a point internal to an area was selected with the pick points option.

```
Command: REGION
Select objects: Specify opposite corner: 5 found
Select objects:
1 loop extracted.
1 Region created.

Command: REGION
Select objects: Specify opposite corner: 3 found
Select objects:
0 loops extracted.
0 Regions created.

Command: REGION
Select objects: Specify opposite corner: 4 found
Select objects:
1 loop extracted.
1 loop rejected.
 Self intersections                        : 1 loop.
0 Regions created.

Command: BOUNDARY
Select internal point: Selecting everything...
Selecting everything visible...
Analyzing the selected data...
Analyzing internal islands...
Select internal point:
1 loop extracted.
1 Region created.
BOUNDARY created 1 region
```

Physical Properties of a Region

One of the advantages of creating a region is the ability to analyze that Region for some basic physical properties. As compared to analytical methods, these make a believer in graphical methods out of most everyone. Once you have created a region, you can use the command MASSPROP to determine a range of physical properties.

```
Command: MASSPROP
Select objects: 1 found
Select objects:
--------------- REGIONS ---------------
Area:                              2.2555
Perimeter:                         6.9710
Bounding box:                 X:   0.5625 -- 2.5625
                              Y:   0.7232 -- 2.9732
Centroid:                     X:   1.7374
                              Y:   2.0654
Moments of inertia:           X:   10.0835
                              Y:   7.2425
Product of inertia:          XY:   8.0487
Radii of gyration:            X:   2.1144
                              Y:   1.7919
Principal moments and X-Y directions about centroid:
                              I:   0.4008 along [0.5929 -0.8053]
                              J:   0.4951 along [0.8053 0.5929]

Write analysis to a file? [Yes/No] <N>:
```

If you are trying to calculate those properties mathematically, then you could use the formulas shown in Figure 10.2. For some region shapes, using the analytical formulas is not too difficult, but in some cases it is almost impossible (actually for some shapes, the analytical solution IS impossible and numerical methods must be used).

These properties can be used to analyze the region and how the part it represents will react to applied forces in the real world. The Area can be used to determine the mass of a piece of sheet metal, the Moments of Inertia tell you how it will react if a moment is applied to initiate rotation about the X or Y axis.

$$A = \int dA \qquad r_x = \sqrt{\frac{I_{yy}}{A}}$$

$$\bar{x} = \frac{\int y\,dA}{\int dA} \qquad r_y = \sqrt{\frac{I_{xx}}{A}}$$

$$\bar{y} = \frac{\int x\,dA}{\int dA} \qquad I_{xxc} = \int (y-\bar{y})^2 dA \qquad I_{yyc} = \int (x-\bar{x})^2 dA$$

$$I_{xx} = \int y^2 dA$$

$$I_{xy} = \int x^2 dA$$

$$I_{xy} = \int xy\,dA$$

Figure 10.2

Location of the Origin

Most of the properties calculated are dependent on the location of the origin. The centroid is certainly origin dependent, the bounding box, the moments of inertia and its derived properties. All of the properties except the Area and the Principle Moments depend on the location of the origin.

Since so many of these properties depend on the actual location of the region, it is very important to draw the region at the correct location on the drawing plane. Sometimes, this is not convenient (suppose you have regions of two gears spinning about different centers) so you need to relocate the origin of the drawing to the region. This is done by defining a new coordinate system, centered at the point you desire.

AutoCAD starts a drawing with a coordinate system called the World Coordinate System (WCS) and all objects and coordinate systems have their roots here. Any coordinate system you create is called a User Coordinate System (UCS). In the next chapter, we will examine many options for creating UCS's, but for regions it is sufficient to simply move the origin. The general command to create user coordinate systems is UCS.

There are many options available to create UCS's (it even has two toolbars devoted to nothing but UCS's) however, the only three options of importance for regions are World, Move, and Z.

World

This restores the coordinate system to the WCS, regardless of what coordinate system is in effect at the time you execute the UCS command.

Move

This will request a new origin location and move the coordinate system origin to that point. The X axis will stay parallel with the former X axis as will the Y.

Z

If you want to rotate the X and Y axes on the drawing, you can use the Z option. It will leave the origin in the same location and spin the XY axes by the specified number of degrees counterclockwise.

One feature we have ignored up to this point is the small icon which notes the direction of the X and Y axes. This is called the Ucsicon and has a few options which should be examined

The Ucsicon

The command to manipulate the icon is UCSICON. The display of the icon can be turned On or Off. For many of the earlier drawings the icon was turned off on the template because you did not need it. Now that we are beginning 3D drawings, the icon becomes important and should always be displayed.

Another option, which many beginning students don't really understand, is the ability to force the icon to track the physical origin of the drawing as opposed to simply noting the directions of the X and Y axes. If you set the icon to Noorigin (No Origin), then the icon will always rest in the lower left corner of the display marking the direction of the X and Y axes. If you set it to Origin, then the icon will TRY to place itself at the actual point 0,0 on the drawing. If this point is off screen or placing the icon there would force any portion of the icon to go off screen, then it will default to the lower left corner like in the Noorigin case.

The final option available is to control the appearance of the icon itself using the Properties option. These will be examined in the next chapter for use in true 3D drawings.

Annotating a Drawing with the Physical Properties

To fully document a drawing it can be dimensioned, but including the physical properties on the printed page is useful for future design work. Placing the annotation on the page involves having AutoCAD calculate the properties and then doing a cut and paste to get the properties from the text screen onto the drawing screen. Once the properties are on the drawing, they can be edited to get to correct text height, or location.

Boolean Operations

The term Boolean operations is used more commonly in mathematics to refer to work done by British Mathematician George Boole who lived in the 1800's. In set theory, the two most common operations are UNION and INTERSECT. We have all done problems where set A contained the numbers {1,2,3,4,5,6} and set B contained the numbers {2, 4, 6, 8, 10} and we were asked to find the intersection between sets A and B $(A \cap B) = \{2, 4, 6\}$ or the union of sets A and B $(A \cup B) = \{1,2,3,4,5,6,8,10\}$. AutoCAD works the same way, but with geometric figures. It also adds the Boolean operator of SUBTRACT such that, using our sets above A subtract B (A – B) = {1, 3, 5} or (B – A) = {8, 10}. (Note that subtract is order dependent.)

Looking at this in a graphical manner, if you have two intersecting regions a Hexagon and a Square, the following Boolean operations are possible:

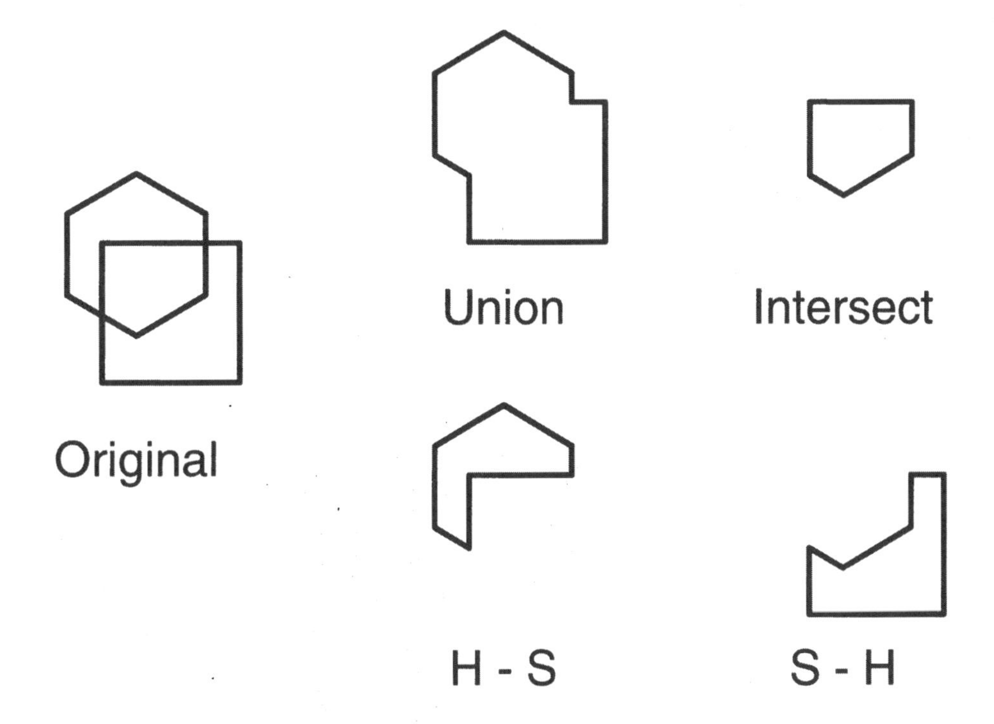

Figure 10.3

Example: Your company plans to make small demo CD's for distribution at an upcoming convention. You need to determine the physical properties for the CD to insure it will work properly in a disk drive. The proposed shape is shown in Figure 10.5.

Figure 10.4

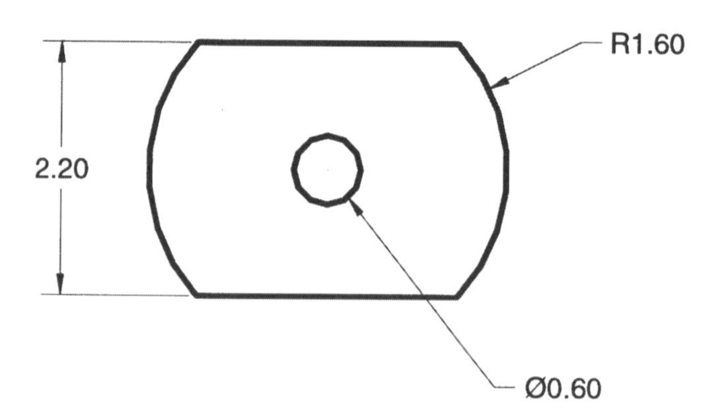

Figure 10.5

Drawing with the center point at 5,5 just to move the drawing away from the border, you initially create the circle of radius 1.6 and the smaller circle.

```
Command: c CIRCLE Specify center point for circle or [3P/2P/Ttr
      (tan tan radius)]: 5,5
Specify radius of circle or [Diameter] <0.5000>: 1.6

Command: c CIRCLE Specify center point for circle or [3P/2P/Ttr
      (tan tan radius)]: 5,5
Specify radius of circle or [Diameter] <1.6000>: .3
```

Next, we create two rectangles to be regions and create the flat sides on the CD.

```
Command: rectang
Specify first corner point or [Chamfer/Elevation/Fillet/
      Thickness/Width]: 0,6.1
Specify other corner point or [Dimensions]: 10,8

Command: rectang
Specify first corner point or [Chamfer/Elevation/Fillet/
      Thickness/Width]: 0,0
Specify other corner point or [Dimensions]: 10,3.9
```

Region the four objects created.

```
Command: region
Select objects: Specify opposite corner: 4 found
Select objects:
4 loops extracted.
4 Regions created.
```

Use the Boolean operation subtract to clip to outer rectangles and the central circle from the larger circle representing the outside of the CD.

```
Command: subtract Select solids and regions to subtract from ..
Select objects: 1 found
Select objects: Select solids and regions to subtract ..
Select objects: 1 found
Select objects: 1 found, 2 total
```

```
Select objects: 1 found, 3 total
Select objects:
```

Move the UCS to the center of the CD, since you want the physical properties as it spins about its center.

```
Command: ucs
Current ucs name: *NO NAME*
Enter an option [New/Move/orthoGraphic/Prev/Restore/Save/Del/
      Apply/?/World] <World>: m
Specify new origin point or [Zdepth]<0,0,0>: 5,5
```

Find the Mass Properties of the CD. Remember to respond to the "Write analysis to a file" prompt.

```
Command: massprop
Select objects: 1 found
Select objects:
 ------------------- REGIONS --------------------

Area:                               6.1546
Perimeter:                          11.3840
Bounding box:                   X:  -1.6000 -- 1.6000
                                Y:  -1.1000 -- 1.1000
Centroid:                       X:  0.0000
                                Y:  0.0000
Moments of inertia:             X:  2.3881
                                Y:  4.6887
Product of inertia:            XY:  0.0000
Radii of gyration:              X:  0.6229
                                Y:  0.8728
Principal moments and X-Y directions about centroid:
                                I:  2.3881 along [1.0000 0.0000]
                                J:  4.6887 along [0.0000 1.0000]

Write analysis to a file? [Yes/No] <N>: N

Command:
```

Highlight the mass properties and copy them to the clipboard. Then press F2 to return to the drawing and paste the copied properties.

```
Command: _pasteclip
```

Modify the text height of the pasted text if necessary.

```
Command:
Command: _properties
```

Move the text from the upper left corner to the correct location on the page. You may have to move the drawing also.

```
Command:
Command: _move 1 found
Specify base point or displacement:
Specify second point of displacement or <use first point as
      displacement>:
```

Your drawing should look like Figure 10.6.

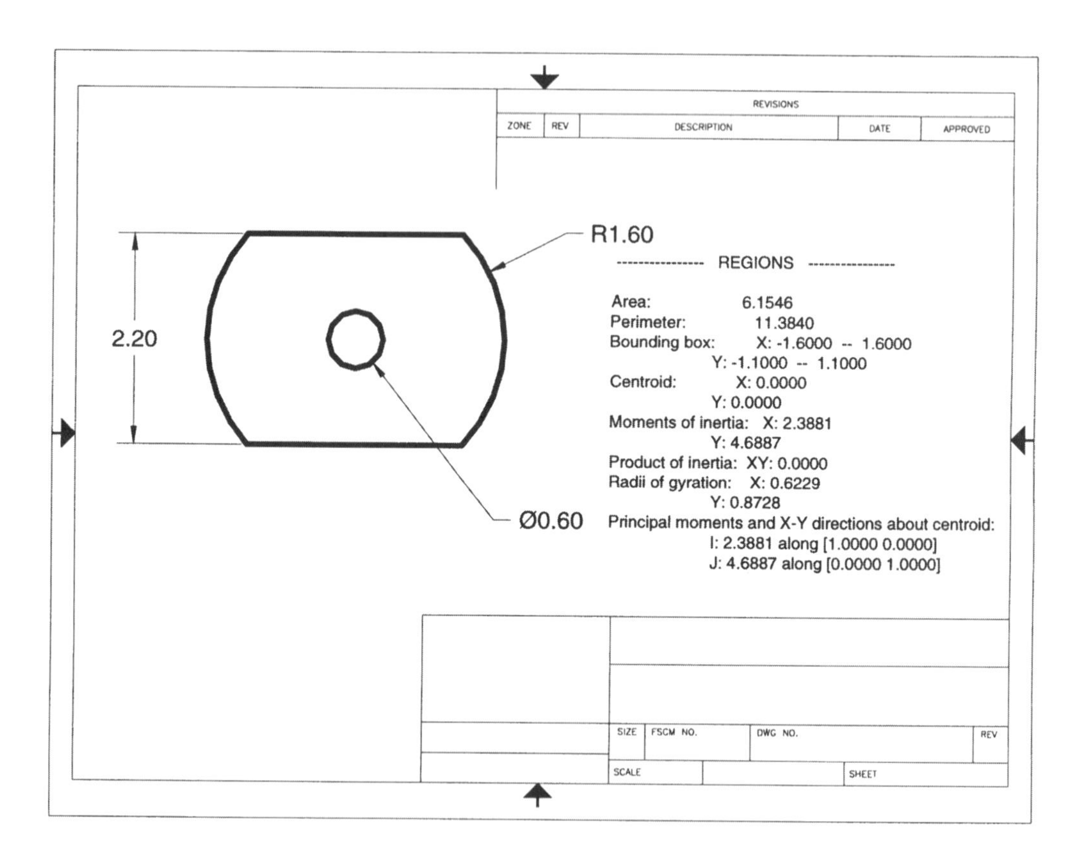

Figure 10.6

3D-Solid Modeling

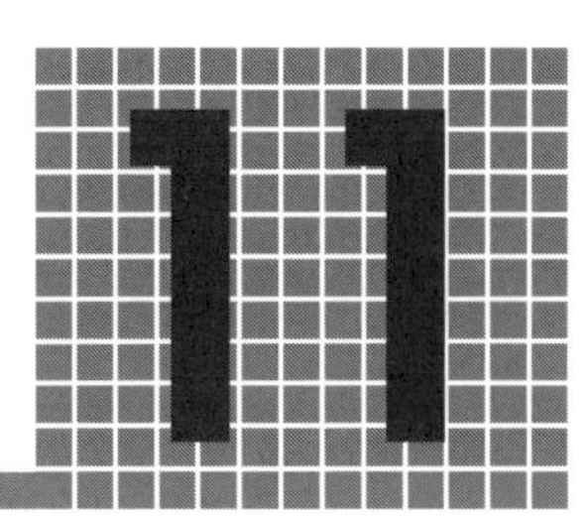

Objectives

1. Know the six solid primitives used by AutoCAD
2. Be able to view a 3D object from different viewpoints
3. Be able to HIDE or SHADE a drawing
4. Be able to create custom shaped solids by EXTRUDING or REVOLVING regions
5. Use Boolean operations on solids
6. Use layout mode to correctly plot 3D drawings
7. Extract mass properties of solids and place them on a drawing
8. Know the various configurations of the UCSICON
9. Be able to establish user coordinate systems on different planes
10. Be able to fillet and chamfer solids
11. Be able to create full sections of solid models using SLICE and SECTION
12. Be able to align multiple solid objects in the same file

While regions are a good introduction to 3D concepts, they do little to really motivate aspiring CAD users to the exciting world of 3D. Once you get a chance to see what is available in 3D and how easy it is to create and display basic objects in 3D, this program called AutoCAD becomes more like a toy than a tool.

Solid Primitives

Many objects in this world can be easily represented by the combination of several basic 3D shapes. AutoCAD supports six basic shapes, called primitives. These are: Box, Sphere, Cylinder, Cone, Wedge, and Torus. These can be most easily drawn by displaying the Solids toolbar as shown in Figure 11.1. By selecting the appropriate icon you can draw the given shape. Figure 11.2 shows the application of each of the primitives and the command stream associated with the figure.

Figure 11.1

```
Command: _box
Specify corner of box or [CEnter] <0,0,0>:
Specify corner or [Cube/Length]:
Specify height: 2
Command:
Command:
Command: _sphere
```

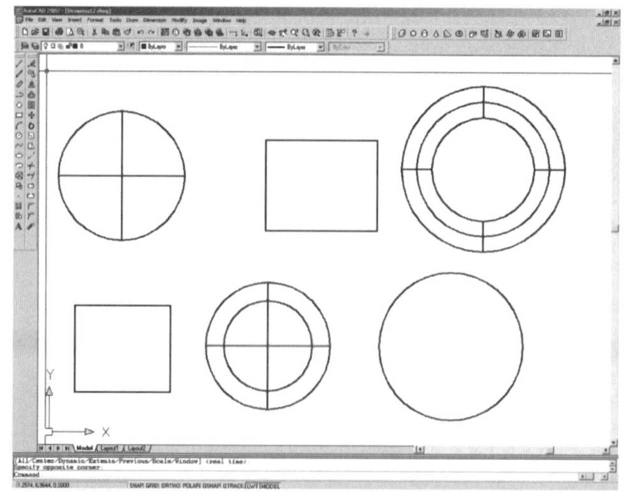

Figure 11.2

```
Current wire frame density: ISOLINES=4
Specify center of sphere <0,0,0>:
Specify radius of sphere or [Diameter]:
Command:
Command:
Command: _cylinder
Current wire frame density: ISOLINES=4
Specify center point for base of cylinder or [Elliptical]
      <0,0,0>:
Specify radius for base of cylinder or [Diameter]:
Specify height of cylinder or [Center of other end]: 2
Command:
Command:
Command: _cone
Current wire frame density: ISOLINES=4
Specify center point for base of cone or [Elliptical] <0,0,0>:
Specify radius for base of cone or [Diameter]:
Specify height of cone or [Apex]: 2
Command:
Command:
Command: _wedge
Specify first corner of wedge or [CEnter] <0,0,0>:
Specify corner or [Cube/Length]:
Specify height: 2
Command:
Command:
Command: _torus
Current wire frame density: ISOLINES=4
Specify center of torus <0,0,0>:
Specify radius of torus or [Diameter]:
Specify radius of tube or [Diameter]: Specify second point:
```

The problem with Figure 11.2 is that it does not look like a 3D drawing. All the objects are shown from a top view, thus the cylinder looks like a circle, the box looks like a rectangle, etc. This is one of the features of AutoCAD which the beginning students needs to get used to. Initially, AutoCAD begins by displaying a top view of the drawing and allowing you to draw on the horizontal projection plane. This can be changed by changing coordinate systems as will be addressed later in the chapter, but it requires action on the user's part. The default setting is to allow construction on the horizontal plane.

While changing coordinate systems to allow construction on different planes is important, a first step is to simply change the viewing direction for the drawing, thus allowing a more 3D appearance of the drawing.

The most flexible and probably easiest way to do this is to use the 3DORBIT command. The icon is located on the Standard toolbar located at the top of the screen. Once selected, the screen will change to display a green circle superimposed on top of the drawing as shown in Figure 11.3.

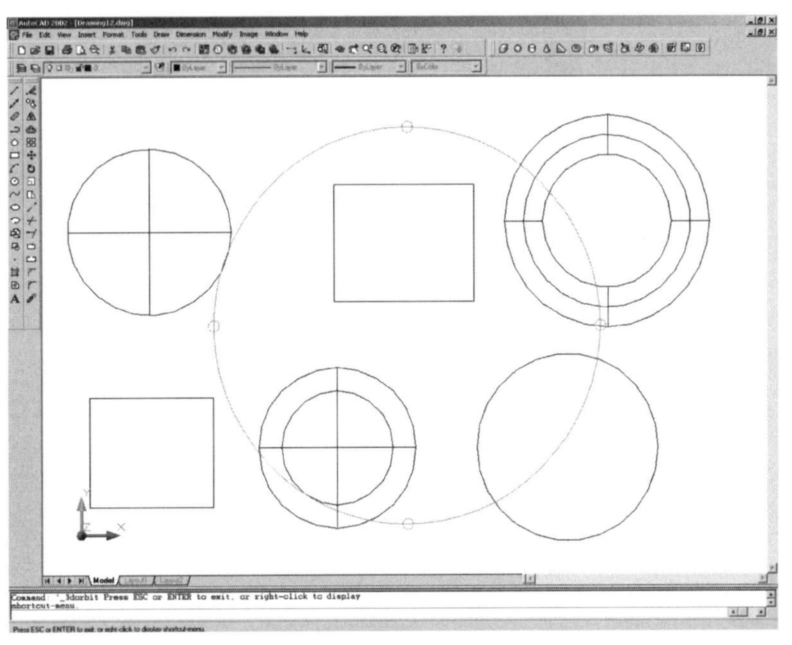

Figure 11.3

By placing the crosshairs (which are now a small strange looking symbol) inside the green circle and holding down the pick button, you can drag the display in 3D movement. This will reveal the true nature of the drawing (Figure 11.4).

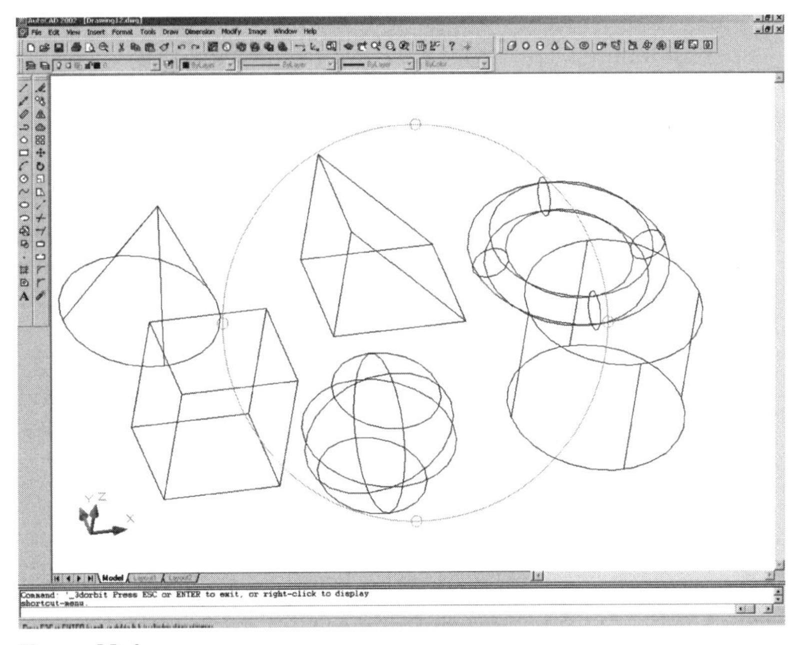

Figure 11.4

Once you get the display the way you want it, press enter and the green circle will disappear.

You can now see all the shapes as 3D objects, however, they are transparent, only the edges (or a few structure lines, in the case of curved surfaces) show up. This is called a wireframe display and is the simplest method of showing 3D objects. If you want to make the objects look solid, you must tell AutoCAD to HIDE the display, thus showing the objects as if they were opaque. A hidden view of the drawing is shown in Figure 11.5.

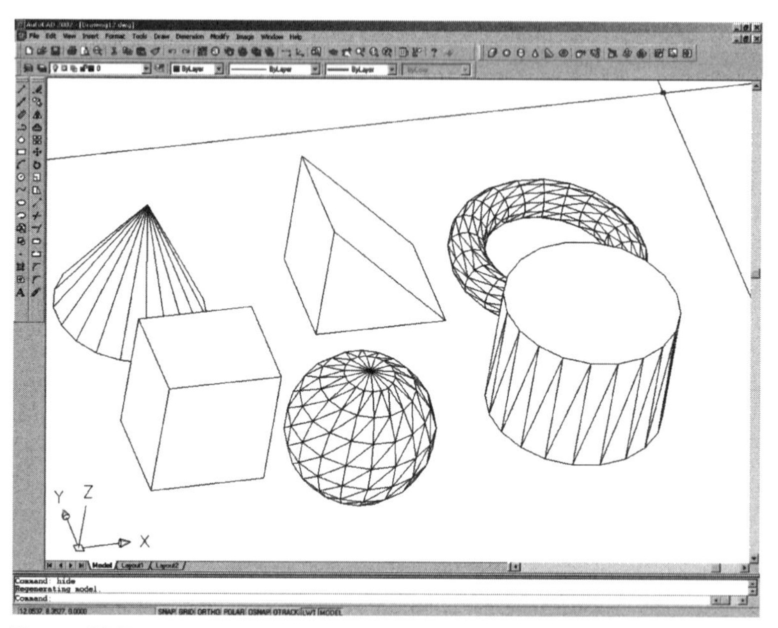

Figure 11.5

If you execute a command that forces the drawing to be re-displayed, then the drawing will not be hidden (examples are 3Dorbit, Regen, and Zoom). To keep the display hidden even when changing viewpoints or zooming, you should use the pulldown menu (Figure 11.6) for view and select Shade. This will give you access to the various possible modes. Once set, these modes will be in effect until they are changed by the user. Any redisplay of the drawing will still be displayed using the current settings.

The effect of each of the modes is shown in Figure 11.7. There is no difference in how the objects are displayed with either the 2D Wireframe or 3D Wireframe options other than line weight settings are not displayed with the 3D option (but since most users have the line weight setting turned off when drawing, this is a minor point).

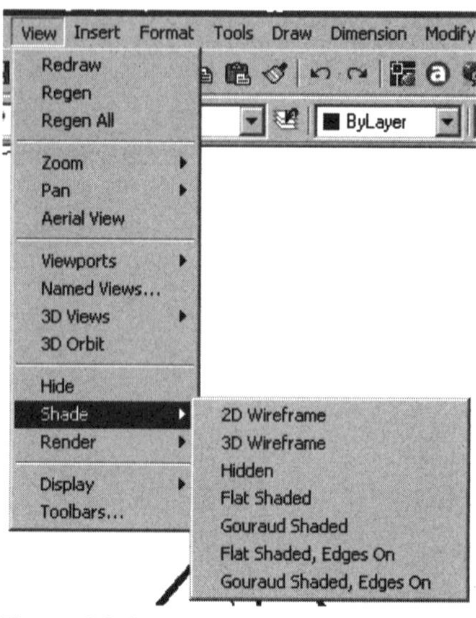

Figure 11.6

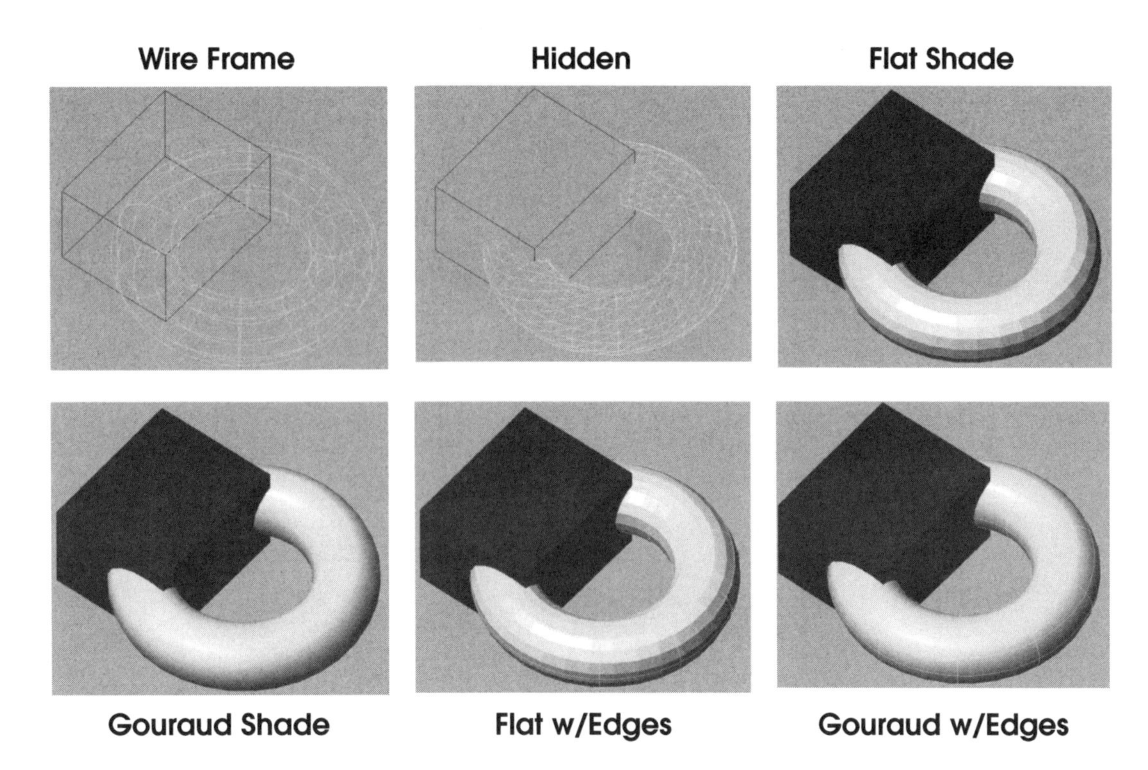

Figure 11.7

Extrusions and Revolutions

While it is possible for many objects to be created strictly from primitive shapes, frequently real world objects have characteristic shapes that need more than simple primitive shapes to develop. AutoCAD supports the ability to use a region as a characteristic shape to define solid volumes by sweeping that shape along a path, either straight line (extrusion) or circular (revolution) to define a solid object. It is also possible to sweep along an irregular path, but this requires the use of user coordinate systems and will be addressed at that time.

When using either the EXTRUDE or REVOLVE command, you must first define the region for your basic shape. This can be any valid region in AutoCAD. Draw the shape with normal objects, create a region from them, perform any Boolean operations you want and get the region into final form, then invoke the EXTRUDE (or REVOLVE) command.

```
Command: _line Specify first point: <Snap on>
Specify next point or [Undo]:
Specify next point or [Undo]:
Specify next point or [Close/Undo]:
Specify next point or [Close/Undo]:
Specify next point or [Close/Undo]:
Command: A ARC Specify start point of arc or [Center]:
Specify end point of arc:
Command: L LINE Specify first point: Length of line:
Specify next point or [Undo]:
```

```
Command: _region
Select objects: Specify
      opposite corner: 6
      found
Select objects:
1 loop extracted.
1 Region created.
Command: extrude
Current wire frame
      density:
      ISOLINES=4
Select objects: 1 found
Select objects:
Specify height of
      extrusion or
      [Path]: 3
Specify angle of taper
      for extrusion <0>:
      10
```

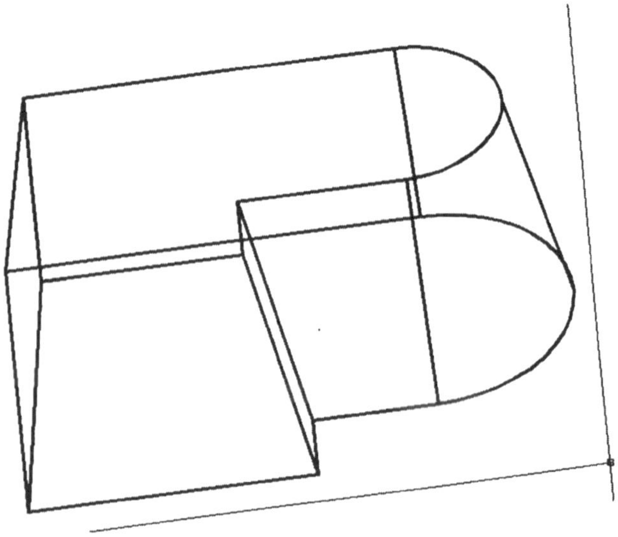

Figure 11.8

```
Command: '_3dorbit Press ESC or ENTER to exit, or right-click to
      display shortcut-menu.
Regenerating model.
```

When you are using the EXTRUDE command, you are asked for 3 items, the region(s) you want to use, the extrusion height, and the taper angle. The height is the total height of the solid object, the angle is the slope from the vertical for each of the sides. A positive height will extrude upward, a negative will go down. A positive angle with slope inwards (as shown in Figure 11.8), a negative will slope outwards.

REVOLVE works in much the same way.

```
Command: polygon Enter number of sides <4>: 3
Specify center of polygon or [Edge]:
Enter an option [Inscribed in circle/Circumscribed about circle]
      <I>: i
Specify radius of circle:
Command: region
Select objects: L 1 found
Select objects:
1 loop extracted.
1 Region created.
Command: revolve
Current wire frame density:
      ISOLINES=4
Select objects: L 1 found
Select objects:
Specify start point for axis of
      revolution or define axis
      by [Object/X (axis)/Y
      (axis)]: x
Specify angle of revolution
      <360>:
Command: '_3dorbit Press ESC or
      ENTER to exit, or right-
      click to display
shortcut menu.
Regenerating model.
```

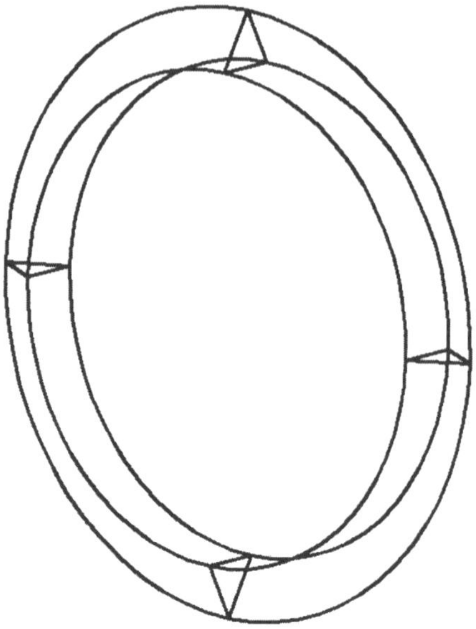

Figure 11.9

This will create a triangular torus as shown in Figure 11.9. Note the REVOLVE command also requires 3 items, the region, the axis of revolution (the X-axis in this case), and the total angle of revolution around that axis.

By combining these two objects, we can examine the effects of Boolean operations on solids.

Using Boolean Operations for Solids

The basic concepts of Boolean operations are the same with solids are they are with regions. In Figure 11.10, we have moved the Extrusion and the Revolution from above to overlap and then performed the various Boolean operations on the pair of objects. Across the top row are the original objects, the union of the two, and the Extrusion subtract the Revolution. The second row is the Revolution subtract the Extrusion and the Intersection of the two.

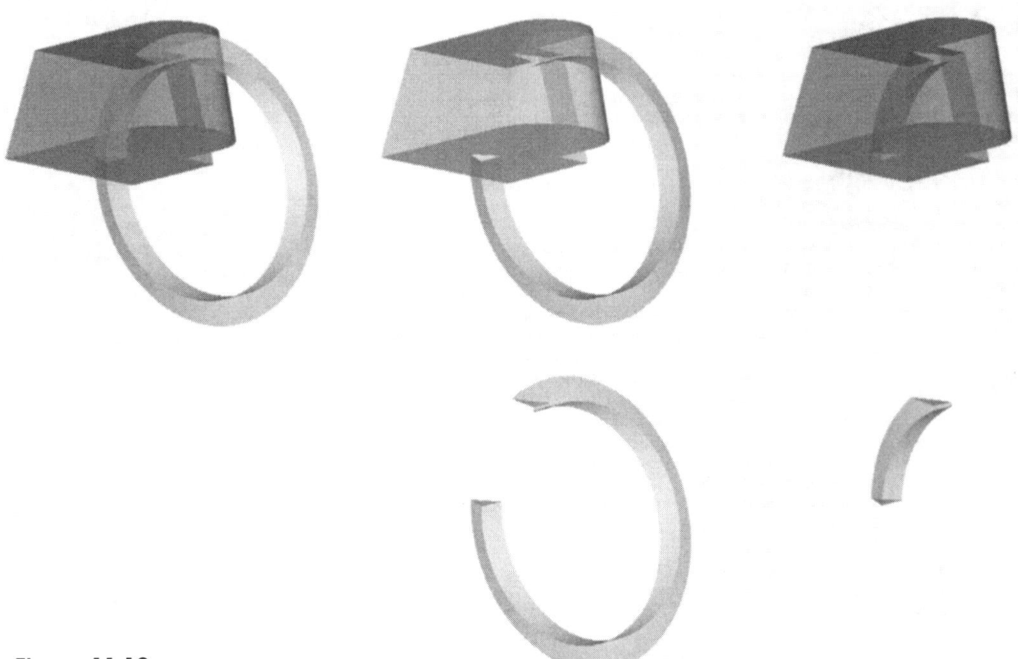

Figure 11.10

Practical Example using Solid Modeling

Create a solid model of a 7/8" Wrench Socket (Figure 11.11). A drawing of the socket is shown in Figure 11.12. Before we even begin to use AutoCAD to draw the solid model of the socket, we must first develop a strategy about how we approach the drawing. AutoCAD always begins a drawing working from the top view, so we should concentrate our attention on that view when considering our approach. Looking at Figure 11.12, all the features in the top view are visible, since the holes get progressively smaller as you go down the part, so when we draw the view we need to extrude things in the negative direction. We will draw the cylinders and the square cutout as primitives and the overlapping hexagons as regions and then extrude them. So the process will be to draw two

Figure 11.11

cylinders, one Ø1.25 with a height of –1.5, the other Ø0.75 with a height of –1.00. The square will be a box 0.5"x 0.5" with a height of –2.00" (the height does not really matter since the box will penetrate out the bottom side anyway, so using 2" will guarantee it clears the bottom). We will then draw two 6 sided polygons, each circumscribed about a Ø7/8" circle, offset by 90 degrees. Region them and extrude them –0.75". We will then subtract the smaller cylinder, the box, and the two hexagonal prisms from the large cylinder to complete the socket. Since all of the dimensions are in multiples of 1/8" we will set our snap to 1/16" (since some are diameter measurements) and our grid to 1/8".

With our strategy laid out, we begin a new drawing and start construction.

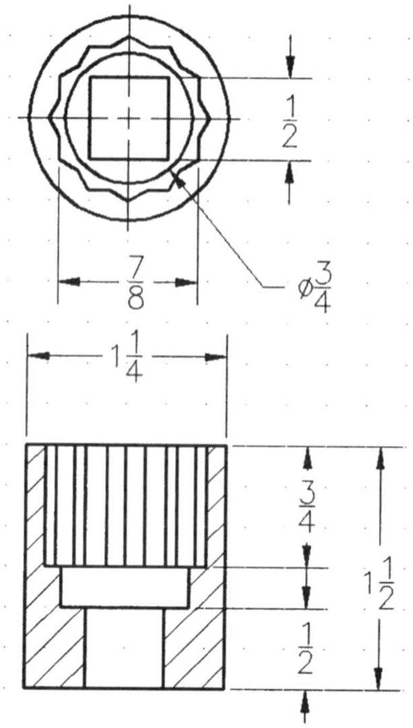

Figure 11.12

```
Command: grid
Specify grid spacing(X) or [ON/OFF/
      Snap/Aspect] <0.5000>: 1/8
Command: snap
Specify snap spacing or [ON/OFF/
      Aspect/Rotate/Style/Type]
      <0.5000>: 1/16
Command: _cylinder
Current wire frame density: ISOLINES=4
Specify center point for base of cylinder or [Elliptical]
      <0,0,0>:
Specify radius for base of cylinder or [Diameter]: d
Specify diameter for base of cylinder: 1.25
Specify height of cylinder or [Center of other end]: -1.5
Command: _cylinder
Current wire frame density: ISOLINES=4
Specify center point for base of cylinder or [Elliptical]
      <0,0,0>:
Specify radius for base of cylinder or [Diameter]: d
Specify diameter for base of cylinder: .75
Specify height of cylinder or [Center of other end]: -1.00
Command: box
Specify corner of box or [CEnter] <0,0,0>:
Specify corner or [Cube/Length]:
Specify height: -2
Command: z ZOOM
Specify corner of window, enter a scale factor (nX or nXP), or
[All/Center/Dynamic/Extents/Previous/Scale/Window] <real time>:
Specify opposite corner:
Command: polygon
Enter number of sides <4>: 6
Specify center of polygon or [Edge]:
Enter an option [Inscribed in circle/Circumscribed about circle]
      <I>: c
Specify radius of circle: {drag to the left}
Command: POLYGON Enter number of sides <6>:
Specify center of polygon or [Edge]:
```

```
Enter an option [Inscribed in circle/Circumscribed about circle]
      <C>: c
Specify radius of circle:{drag to the top}
```

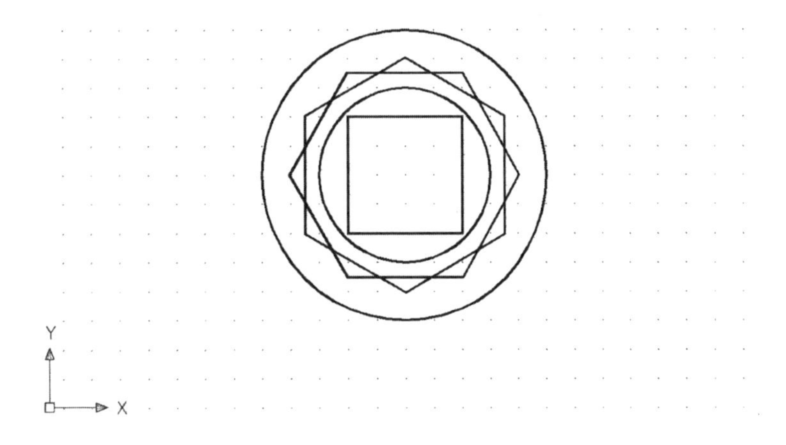

Figure 11.13

```
Command: _region
Select objects: 1 found
Select objects: 1 found, 2 total
Select objects:
2 loops extracted.
2 Regions created.
Command: _extrude
Current wire frame density: ISOLINES=4
Select objects: 1 found
Select objects: 1 found, 2 total
Select objects:
Specify height of extrusion or [Path]: -.75
Specify angle of taper for extrusion <0>:
```

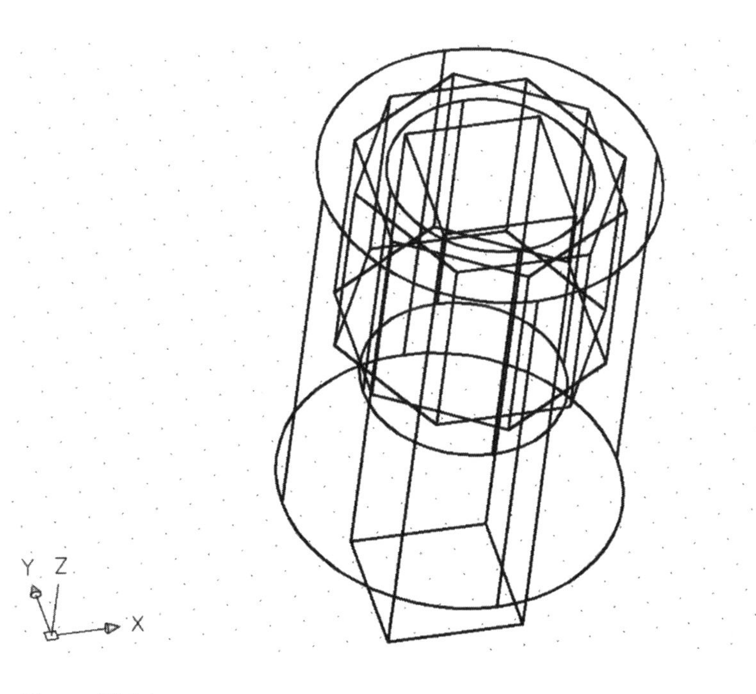

Figure 11.14

```
Command: '_3dorbit Press ESC or ENTER to exit, or right-click to
      display shortcut-menu.
Regenerating model.
Command: _extrude
Current wire frame density: ISOLINES=4
Select objects: 1 found
Select objects: 1 found, 2 total
Select objects:
Specify height of extrusion or [Path]: -.75
Specify angle of taper for extrusion <0>:
Command: '_3dorbit Press ESC or ENTER to exit, or right-click to
      display shortcut-menu.
Regenerating model.
Command: subtract Select solids and regions to subtract from ..
Select objects: 1 found {select Larger Cylinder}
Select objects: Select solids and regions to subtract ..
Select objects: <Snap off> 1 found {Select Smaller Cylinder}
Select objects: 1 found, 2 total {Select Box}
Select objects: 1 found, 3 total {select on Hexagon}
Select objects: 1 found, 4 total {Select other Hexagon}
Select objects:
Command: '_3dorbit Press ESC or ENTER to exit, or right-click to
      display shortcut-menu.
Regenerating model.
Command: hide
Regenerating model.
```

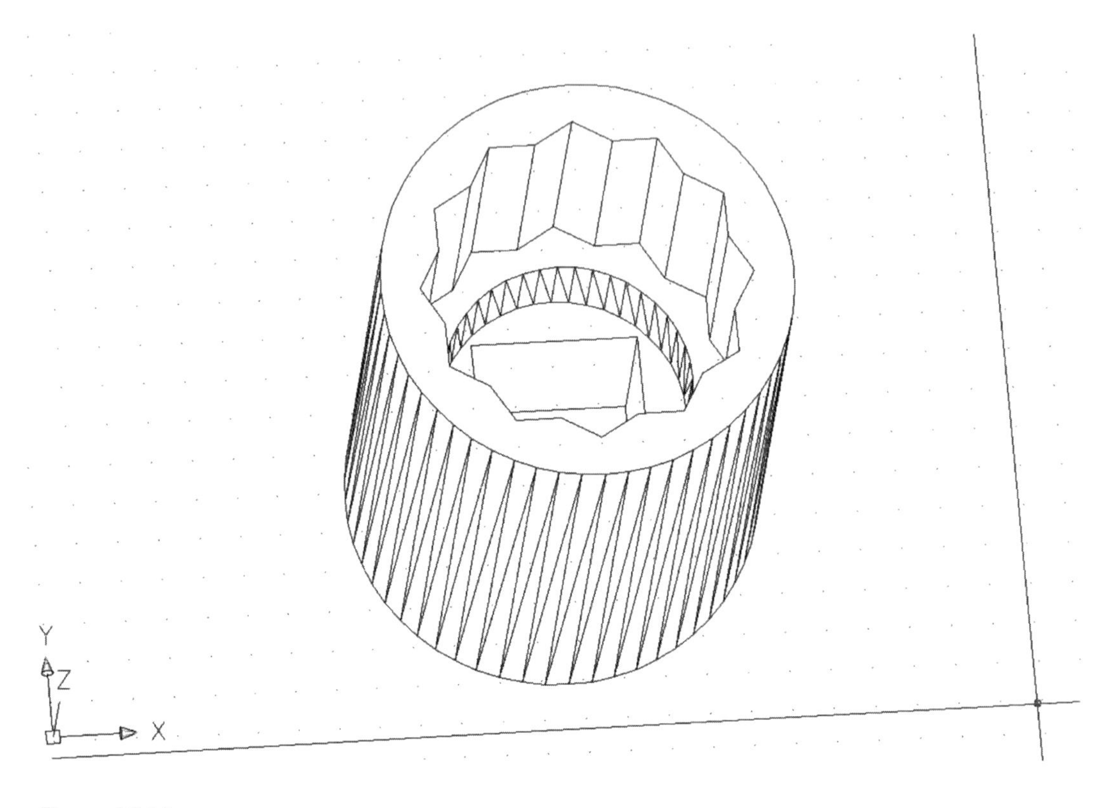

Figure 11.15

Plotting of 3D Drawings

If you wanted a hard copy of the socket we just drew, you can plot it like any other drawing in AutoCAD, however, if you wanted to make it look like a formal engineering drawing, complete with border and title block, the procedure is a bit more complicated. We have to use a new concept called layout mode.

To activate Layout Mode, select one of the layout tabs shown at the bottom of the drawing area. If you started a drawing from scratch, then you will be asked to set up your page. If you are using one of the templates from the companion workbook by Dr. Vinson, then "Layout 1" should be configured already. In Layout mode, there are two (or more) distinct drawing areas superimposed on one another. One of them (called paperspace) represents the actual plotted page and has information like the title block, border, and any administrative notes on it. The other is your original drawing (called modelspace). On the row of modes at the bottom of your drawing you will see one labeled either Paper or Model. Clicking on this button will switch you from one space to the other. You may also double click within either to activate it on screen.

When you are in modelspace, you can edit your object, rotate the viewpoint, or make other modifications to the appearance of the object. Your crosshairs only exist within the window to the object. While you are in paperspace, you cannot edit the object. You can add notes to the drawing or complete the title block.

One convenient metaphor to use when considering the relationship between paperspace and modelspace is to consider paperspace as one side of a cardboard box. Inside the box is the object you are actually trying to draw. On the side of the box, you can write anything you want about the object, but if you want to actually see the object, you must cut a hole in the cardboard. This hole is the viewport (also called an MVIEW or model view) to allow viewing of the object. If you reach through the hole, you can rotate the model or even drill a hole in it.

The one other major property of the actual viewport is important when plotting 3D drawings. When AutoCAD prints a drawing, it basically re-draws it to the printer. The problem this causes when working with 3D models is that they will always unhide when they are printed. This can be changed for an individual viewport by changing the status of the Hideplot option for the view. The Hideplot option can be accessed from the MVIEW command. It can be either turned on or off for each viewport on a drawing (yes, you can have more than one viewport).

For our wrench socket the command stream would be:

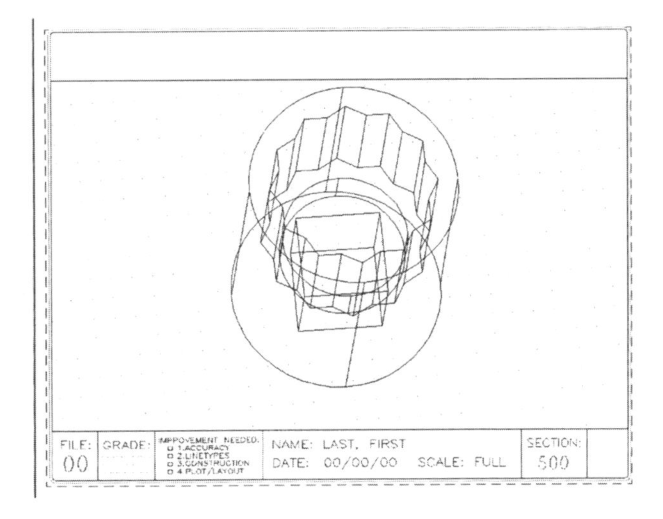

Figure 11.16

```
Command: mview
Specify corner of viewport or [ON/OFF/Fit/Hideplot/Lock/Object/
      Polygonal/Restore/2/3/4] <Fit>: hideplot
Hidden line removal for plotting [ON/OFF]: on
Select objects: 1 found {select edge of viewport, not the
      socket}
Select objects:
```

It is sometimes difficult to find the exact edge of the viewport, so one useful technique is to use the "all" option for selecting the viewport. That will select everything in the drawing, and it will reject anything that is not a viewport, thus you can select the viewport without worrying about finding its edge.

```
Command: mview
Specify corner of viewport or [ON/OFF/Fit/Hideplot/Lock/Object/
      Polygonal/Restore/2/3/4] <Fit>: hideplot
Hidden line removal for plotting [ON/OFF]: on
Select objects: all
42 found
40 were not a viewport.
1 was the paper space viewport.
Select objects:
```

As long as you want all your viewports to be hidden, this method will always work.

Physical Properties

Now that we have a solid model of the Wrench Socket, we would like to have AutoCAD help with the engineering analysis of this object. How much does it weigh? If you put it on the end of an impact wrench, how much torque does it take to accelerate it to 1000 rpm in 1 second? Where is the center of mass? All of these questions deal with the physical properties of the object. These can be determined by using the MASSPROP command. For the socket wrench, assuming the cylinder began at the point 0,0,0, we get the following output from the MASSPROP command:

```
Command: massprop
Select objects: 1 found
Select objects:
 --------------------- SOLIDS ---------------------

Mass:                        1.0723
Volume:                      1.0723
Bounding box:            X:  -0.6250 -- 0.6250
                         Y:  -0.6250 -- 0.6250
                         Z:  -1.5000 -- 0.0000
Centroid:                X:  0.0000
                         Y:  0.0000
                         Z:  -0.8652
Moments of inertia:      X:  1.1406
                         Y:  1.1406
                         Z:  0.2860
Products of inertia:    XY:  0.0000
                        YZ:  0.0000
                        ZX:  0.0000
```

```
Radii of gyration:          X:  1.0313
                            Y:  1.0313
                            Z:  0.5165
Principal moments and X-Y-Z directions about centroid:
                            I:  0.3378 along [1.0000 0.0000 0.0000]
                            J:  0.3378 along [0.0000 1.0000 0.0000]
                            K:  0.2860 along [0.0000 0.0000 1.0000]
```

When calculating these properties, AutoCAD makes a few assumptions about the object. First, it is of uniform density, thus the centroid and the center of mass are identical. Second, the density of the object is 1 regardless of the units associated with the drawing or the density. Once you realize these assumptions these properties become useful.

The Mass of the object will be the same as the volume, since the density is one. However, if you know the density (tool steel has a density of 0.278 lb_m/in^3), then the actual mass of the part can be found by multiplying the given mass by the density. A word of warning is to make sure the length units in the density you are using match the length units for your drawing. Using this value, the socket would actually weigh 0.298 pounds.

Volume is reported in cubic whatever units the drawing was done in. Since our units were inches, then the volume is 1.0723 in^3.

Bounding Box is just a measure of the physical area of space that the object occupies. The units here are in base length units, inches in this case.

Centroid is the point (x, y, and z) where the object will balance. Again, the units are in base length units.

Moments of inertia represent the ability of an object to resist changes in angular velocity when a torque is applied. It is very analogous to the mass of an object representing the ability of an object to resist acceleration when a force is applied. This property is also affected by the density of the object, since a heavier object is harder to spin than a lighter one. It is also affected by the location of the origin, since the farther away from the axis of rotation a weight is, the harder it is to spin. To find an actual physical value for this socket, we need to multiply the values reported by AutoCAD by the density as was done for mass. This means that the moment of inertia for our socket about the Z axis (which is the axis we would try to rotate around) is 0.0795 $lb_m \cdot in^2$.

Product of inertia are similar to Moments of inertia and are used to determine the forces causing motion of the object. These are in the same units as Moments of inertia.

Radius of Gyration is very similar to centroid, however, it deals with the location of a point mass which would yield the same moment of inertia and the entire object. This does not depend on density and has units of length.

Principle Moments are moments of inertia about axes which run through the centroid of the object. One of these axes causes the object to have the highest moment of inertia possible when spun about an axis passing through this point, another causes the lowest moment of inertia. These two axes are orthogonal to one another and the third is orthogonal to these two. To get physical values from this you must also multiply the values reported by AutoCAD by the density of the object.

Annotating a Drawing with the Physical Properties

Having determined the physical properties, it would be nice to annotate the drawing with these properties to facilitate further engineering analysis. The basic method is very straight forward, however, there are several details which cause much grief to students (and professionals) if things are not done correctly. The basic method is a cut and paste from the command stream window to the actual drawing.

The correct procedure is to:

1. Determine the mass properties using the MASSPROP command.
2. Complete the command and return to the command prompt.
3. Highlight the properties you wish to place on the drawing and press "control-C".
4. Return to the drawing and make sure you are in paperspace. Not being in paperspace is the number one problem when placing mass properties on a drawing.
5. Press "control-V" to paste the properties.
6. They will appear in the upper left corner of the screen. Use Modify properties to set the text height to a correct value and then, use the move command to position them correctly on the drawing.
7. If they disappear behind the viewport, go into modelspace and change the shade setting back to 2D Wireframe. Other modes create an opaque background and will not display the paperspace objects which appear behind them.
8. Go into Modelspace and pan or zoom the object to a location where it will not conflict with the mass properties.
9. Use DDEDIT to edit the properties and change the mass, moments of inertia, products of inertia, and principle moments to real world units as well as apply correct units to the other properties. You might also change the font for the text to a monospaced font such as Courier to make the columns line up correctly.

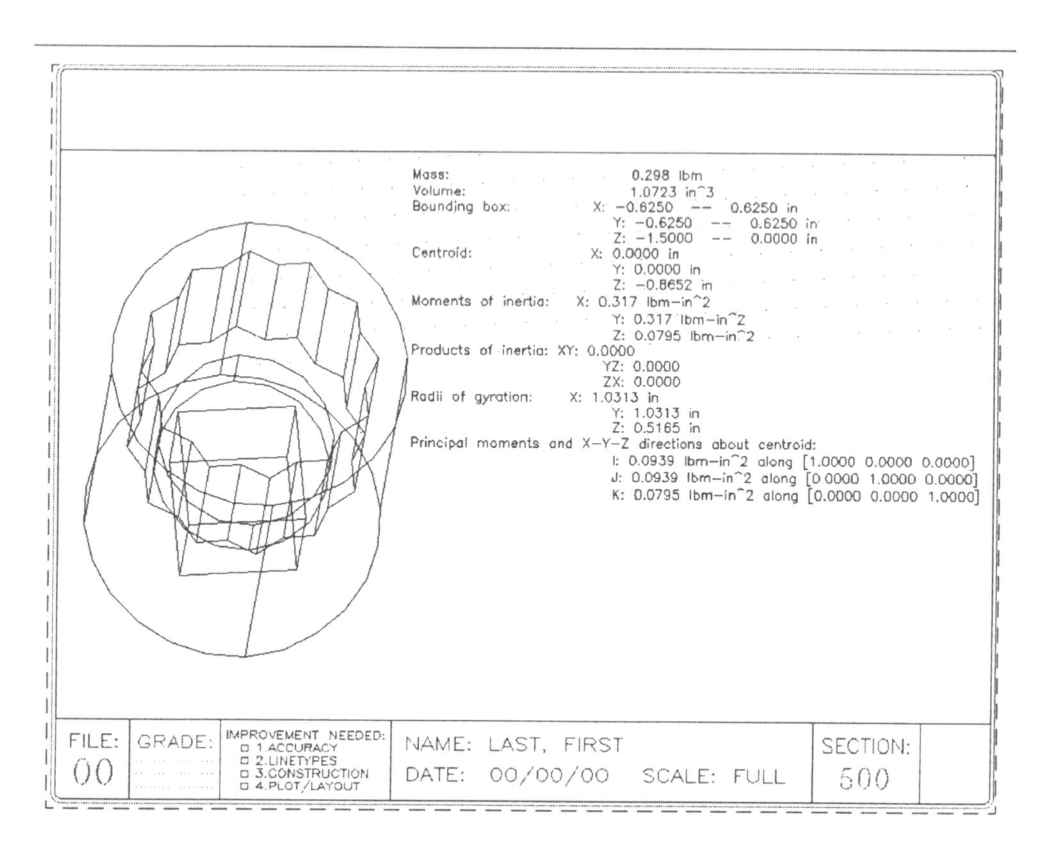

Figure 11.17

Coordinate Systems

There is one major area of 3D drawings that has not yet been discussed. So far, we have limited ourselves to extrusions in one direction, vertically. The real world does not put such limitations on objects. AutoCAD accounts for this by allowing the user to define a coordinate system other than the default which is orientated in any direction and located at any point, as long as it is still orthogonal and right handed.

Once you have established a new coordinate system, AutoCAD will reference this new XY plane as the default drawing surface for construction. All mouse input will be on this new XY plane. You will be able to draw features on sloped surfaces, cut holes in objects at odd angles, construct objects which are not aligned with the world vertical axis. It really opens the world of 3D construction.

Viewing the UCS Icon

Before actually getting into a discussion of defining new coordinate systems, there is one important auxiliary topic dealing with coordinate systems, the UCSICON. The UCS icon is a visual reminder of which way the coordinate system is oriented. It can take several forms and has a few options which are important to understand, lest you get lost in space and not know which way is up.

The first important option is whether it is displayed on the drawing or not. It is possible to turn the icon off. When drawing is strictly 2D, this may be desirable since the icon can clutter the drawing, however in 3D, the icon can give you so much information that turning it on is almost essential, especially to a beginning student. The icon can be activated or deactivated by using the View pulldown menu as shown in Figure 11.18, or the command UCSICON will offer options of ON or OFF.

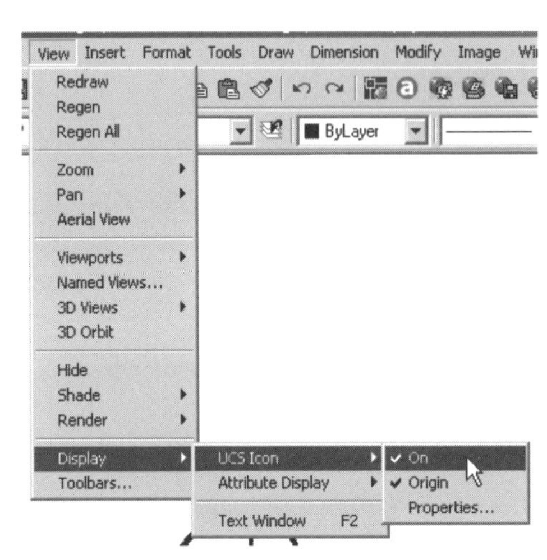

Figure 11.18

Once it is activated, you have the choice of locating the icon at the true origin of the drawing or having the location default to the lower left corner of the screen. While in the lower left corner of the screen, the icon gives information about the orientation of the axes (their direction), but not positional information. This is a good compromise between turning the icon off or leaving it at the true origin if the icon is getting in the way of seeing what you are doing on the drawing, however for most drawings it is best to position the icon at the actual origin of the drawing. If the origin is not visible on the screen (or is very close to the edge), the icon will drop back to the lower left corner until the origin re-appears.

The other option on the menu is Properties. This will activate a dialog box as shown in Figure 11.19. The significant option is the display of the icon as a 2D icon or as a 3D icon. A comparison of the icons is shown in Figure 11.20. The rest of this book will assume that the icon is set to 3D mode.

There is a lot of information embedded within the icon. You can tell whether you are in the World Coordinate System or a User Coordinate System. You can tell if it is located at the actual origin or is just located in the lower left corner of the drawing. The four icons shown in Figure 11.21 all appear to be the same at first glance, but if

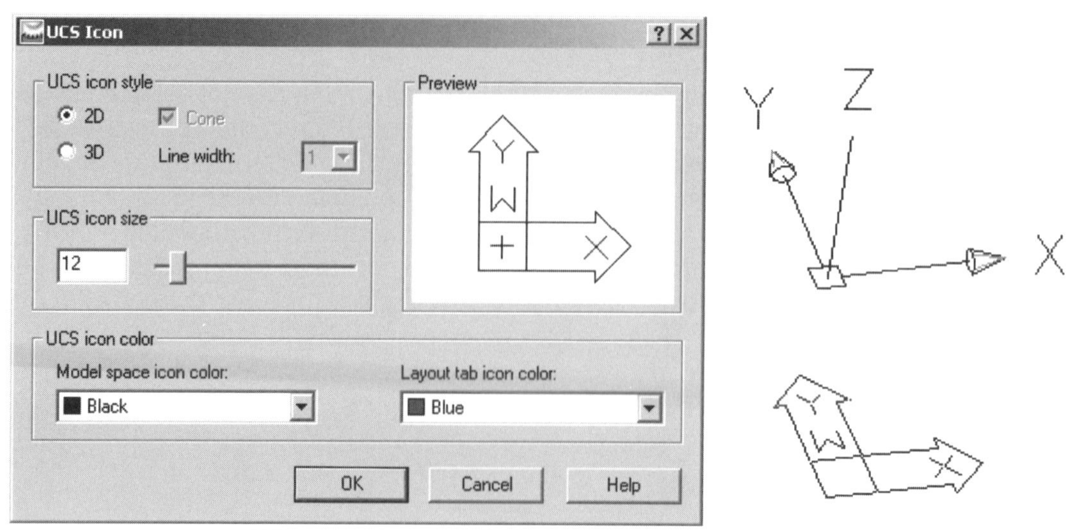

Figure 11.19

Figure 11.20

you look closely at the intersection of the axes, you will see slightly different symbols present. If you see a square located around the intersection, then the icon represents the World Coordinate System (top two icons). If you see a small plus sign, then the icon is located at the true origin of the drawing (left two icons).

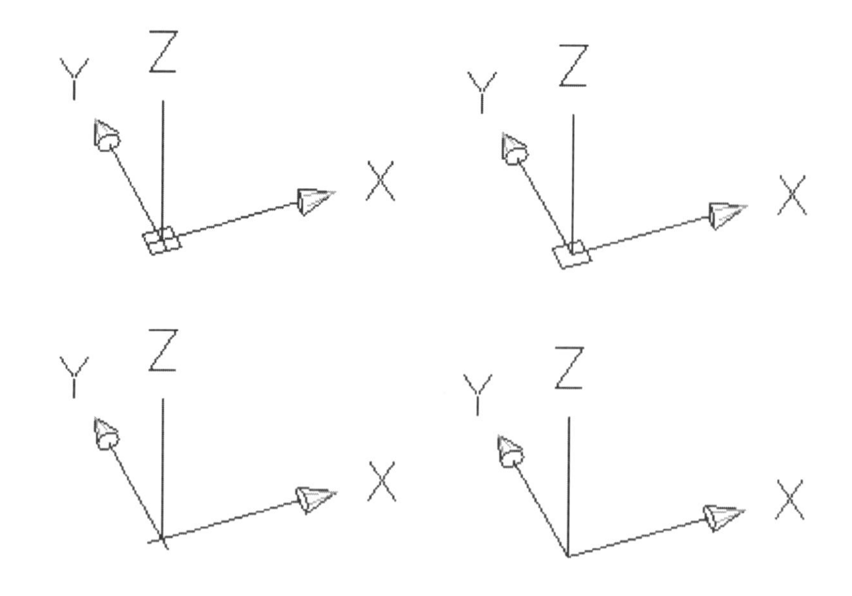

Figure 11.21

Modifying the Coordinate System

The initial default coordinate system is called the World Coordinate System (WCS) and is the base reference for all things in AutoCAD. It cannot be changed, moved, deleted, or modified in any way. It is absolutely fixed and will remain there forever. This is a comforting thing, since you will always have a fixed safety net to return to if you get lost in manipulating coordinate systems. In contract to the WCS, any coordinate system which you define will be called a User Coordinate System (UCS).

In the chapter about regions, we discussed moving the origin of a coordinate system to a new location by using UCS-Move. We now want to extend that discussion to include re-orienting the axes. While doing this, it is possible to leave the origin at the same point and simply rotate the axes around that point, or you can translate and rotate the coordinate system concurrently.

Pure Rotation of the Axes

The UCS command has 4 options which simply rotate the coordinate system. These are X, Y, Z, and View. The first three options simply leave one axis fixed and rotate the other two around the fixed one. Thus, you can rotate the coordinate system 90 degrees around the X axis. Doing this would change the ucsicon from the left icon (note the world coordinate system) to the center icon (a user coordinate system). All construction would now be on a frontal plane, not the initial horizontal plane. To go to a profile plane, you could now rotate 90 degrees about the Y axis yielding the right most icon. Note that the Z axis is shown as a dashed line, indicating that you are viewing the bottom side of the XY plane.

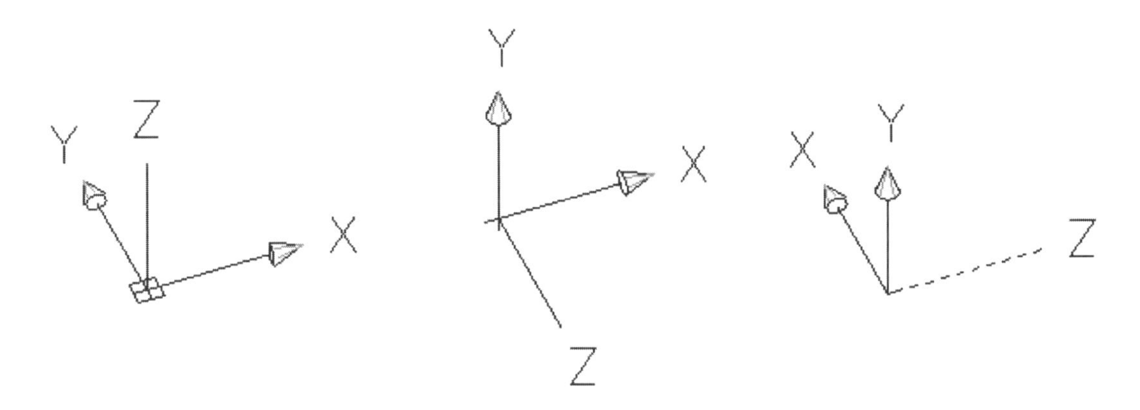

Figure 11.22

The final option under pure rotation is View. This will align the X and Y axes with the edges of the screen. It is marginally useful for annotating a drawing before plotting, since all the text will be readable on the screen, however the better option is to use layout mode for this anyway.

Combination Rotation and Translation of the Axes

While rotation of a fixed number of degrees about a given axis is very useful at times, so is the ability to both move and rotate the coordinate system concurrently. This is most manifest when you have an existing object which you want to locate the XY plane on a surface to add features to the object. The two options of the UCS command which do this for you are Face and 3-Point. If you want to align your system with the side of a solid model, then the Face option is the simplest method to use. You will be asked to select the face of a solid by selecting one of the edges of the solid. Since each edge borders two faces, the next option will toggle between the faces bounded by the selected edge. The origin of the new coordinate system will be the closest corner to the selected point. Xflip and Yflip will rotate the system 180 degrees about that axis to help align the system correctly.

```
Command: ucs
Current ucs name: *WORLD*
Enter an option [New/Move/orthoGraphic/Prev/Restore/Save/Del/
      Apply/?/World] <World>: new
```

```
Specify origin of new UCS or [ZAxis/3point/OBject/Face/View/X/Y/
      Z] <0,0,0>: face
Select face of solid object:
Enter an option [Next/Xflip/Yflip] <accept>: next
Enter an option [Next/Xflip/Yflip] <accept>: X
Enter an option [Next/Xflip/Yflip] <accept>: Y
Enter an option [Next/Xflip/Yflip] <accept>:
```

The 3-Point option is used when no solid face is available. All it requires is 3 non-colinear points to define the XY plane. From this it will determine the Z-axis. The 3 points are, in order, the new origin, a point of the positive X axis, and a point in positive Y space on the XY plane. Using object snap is essential with this command, since the points will not all lie on the current XY plane (otherwise, why are you changing the coordinate system).

```
Command: UCS
Current ucs name: *NO NAME*
Enter an option [New/Move/orthoGraphic/Prev/Restore/Save/Del/
      Apply/?/World] <World>: new
Specify origin of new UCS or [ZAxis/3point/OBject/Face/View/X/Y/
      Z] <0,0,0>: 3point
Specify new origin point <0,0,0>: end of
Specify point on positive portion of X-axis
      <3.3180,6.6365,0.0824>: mid of
Specify point on positive-Y portion of the UCS XY plane
      <1.3834,6.9923,0.0824>: end of
```

The final option of the UCS command we are going to address is the option "World," which simply restores the world coordinate system, regardless of where you are or what coordinate systems you have defined before.

Finishing Touches on Solids

Since most objects in the real world do not have many sharp edges, you will probably need to fillet or round most of the edges of your solid model. The command for this is the same as it was in chapter 6 for rounding the edges of 2D drawings, but the method of application is different. The command is FILLET. The difference is that you are now dealing with solid objects and can clearly specify exactly which edges you want to fillet, not by inferring the edge by picking two adjacent objects as you did in 2D, but by simply selecting the edge you want to fillet. If you have multiple edges you want to fillet, you may continue to select edges until you get all the edges you want to fillet selected.

```
Command: FILLET
Current settings: Mode = TRIM, Radius = 0.5000
Select first object or [Polyline/Radius/Trim]:
Enter fillet radius <0.5000>: .1
Select an edge or [Chain/Radius]:
Select an edge or [Chain/Radius]:
Select an edge or [Chain/Radius]:
3 edge(s) selected for fillet.
```

```
Command: FILLET
Current settings: Mode = TRIM, Radius = 0.1000
Select first object or [Polyline/Radius/Trim]:
Enter fillet radius <0.1000>:
Select an edge or [Chain/Radius]:
Select an edge or [Chain/Radius]:
Select an edge or [Chain/Radius]:
Select an edge or [Chain/Radius]:
Select an edge or [Chain/Radius]:
5 edge(s) selected for fillet.
Modeling Operation Error:
 Geometry or topology at end of blend too complex.
Failed to perform blend.
Failure while filleting.
```

Frequently, you will get an error message similar to the one shown above. This could be due to a great number of things. Try selecting the edges in a different sequence, or performing sequential fillet commands, rather than trying to select all of the edges at once. AutoCAD is very sensitive to sequence and you can frequently work around this error by changing how you select the edges.

Creating a Sectional View of a Solid Model

When printing solid models, you will often want to show some internal detail. This is the same concept as creating a full section (or other type) in conventional orthographic project, but applied to 3D. AutoCAD supports this type of drawing by using 2 commands to operate on a 3D solid model: SLICE and SECTION. The command sequence is similar between the two, but the effects are different.

Both require that you define a plane which intersects the solid model. This can be done by selecting 3 points, by selecting one of the principle planes (XY, YZ, ZX) for the new plane to be parallel to and a point, or several other methods. Once you have the plane defined, the operations diverge.

Slice will actually separate the solid into two different parts, cut along the plane defined and then ask you which portion to keep. You can choose either one or both. If you select both, then you will have two separate solids in your drawing.

Section does nothing to modify the solid itself, but will draw a region in your drawing which is the intersection between the defined plane and the solid. This region is independent of the original solid and can be moved, extruded, or whatever. It can also be crosshatched using the BHATCH command.

One major peculiarity of the BHATCH command is that all crosshatching it creates will always be on the XY plane of the current coordinate system. Thus, before you try to crosshatch a region, you must first establish a coordinate system on that region. Using the 3Point option is most appropriate since a region is not a 3D solid and the face option will not work.

The steps to create a sectioned solid model are:

- ❒ Draw the solid model
- ❒ Extract the mass properties (if needed)
- ❒ Move the UCS in place so that the XY plane is along the desired cutting plane
- ❒ Change layer to HATCH

- ❒ Use the SECTION command along the XY plane
- ❒ Use the SLICE command at the same location
- ❒ Turn off the model layer
- ❒ Use BHATCH to crosshatch the area
- ❒ Turn the model back on

In a pure sense, some of these steps are not strictly necessary, however, AutoCAD will occasionally do something strange if these steps are not followed. Following these steps, in order, to the letter will always work.

Aligning of 3D Drawings

If you are working on a design which has multiple parts to it, aligning them in 3D space can be a difficult proposition. It is much easier to draw each part in its own file and then insert the files into an assembly drawing to show the entire design. The problem is that each part comes in relative to the coordinate system in effect when that file was last saved and the coordinate system in the current assembly drawing.

Rather than worrying about this, it is easier to just insert all the parts into one file and then worry about aligning them. To do this in AutoCAD, you should use the ALIGN command. Align combines 3D move and 3D rotate into a single command based on locating points on both objects and having AutoCAD align those points by moving and rotating one of the objects.

AutoCAD requires either 2 or 3 sets of points to perform the alignment. As you select your point sets (source and destination), AutoCAD will draw temporary lines to remind you of what you have done. The first source points will be placed directly on the first destination point in all cases. The second source point will be placed such that the line from first source point to second source point will be collinear with the same line at the destination. If all you want is axial alignment, then all you need will be these two sets of points and you may press enter rather than selecting a third set of points. If you do this, you will be asked about scaling the object to make it fit exactly. Replying yes to this will enlarge or shrink the object such that the second points will also lie directly on top of each other.

The third set of points is the most difficult to describe in words. Once you have two sets of points and have axial alignment, the final set simply controls rotation about that axis. The third source point will move to make the plane contained by the 3 source points be coplanar with the 3 destination points' plane.

```
Command: align

Initializing...
Select objects: {select the object you want to move} 1 found
Select objects:
Specify first source point: end of
Specify first destination point: end of
Specify second source point: end of
Specify second destination point: end of
Specify third source point or <continue>: end of
Specify third destination point: end of
```

As with 3Point coordinate systems, object snap is essential since you are dealing with points which lie on different planes in 3D space. One common error is to select both objects you want to align. Upon completion of the command, the source object moves into place, but the destination object will move out of the way following the same movement and rotation of the source. If this happens, just undo the command and try again.

AutoCAD Command Aliases

Alias	Command	Alias	Command
3A	3DARRAY	DDI	DIMDIAMETER
3DO	3DORBIT	DED	DIMEDIT
3F	3DFACE	DI	DIST
3P	3DPOLY	DIV	DIVIDE
A	ARC	DLI	DIMLINEAR
ADC	ADCENTER	DO	DONUT
AA	AREA	DOR	DIMORDINATE
AL	ALIGN	DOV	DIMOVERRIDE
AP	APPLOAD	DR	DRAWORDER
AR	ARRAY	DRA	DIMRADIUS
-AR	-ARRAY	DRE	DIMREASSOCIATE
ATT	ATTDEF	DS	DSETTINGS
-ATT	-ATTDEF	DST	DIMSTYLE
ATE	ATTEDIT	DT	DTEXT
-ATE	-ATTEDIT	DV	DVIEW
ATTE	-ATTEDIT	E	ERASE
B	BLOCK	ED	DDEDIT
-B	-BLOCK	EL	ELLIPSE
BH	BHATCH	EX	EXTEND
BO	BOUNDARY	EXIT	QUIT
-BO	-BOUNDARY	EXP	EXPORT
BR	BREAK	EXT	EXTRUDE
C	CIRCLE	F	FILLET
CH	PROPERTIES	FI	FILTER
-CH	CHANGE	G	GROUP
CHA	CHAMFER	-G	-GROUP
COL	COLOR	GR	DDGRIPS
COLOUR	COLOR	H	BHATCH
CO	COPY	-H	HATCH
D	DIMSTYLE	HE	HATCHEDIT
DAL	DIMALIGNED	HI	HIDE
DAN	DIMANGULAR	I	INSERT
DBA	DIMBASELINE	-I	-INSERT
DBC	DBCONNECT	IAD	IMAGEADJUST
DCE	DIMCENTER	IAT	IMAGEATTACH
DCO	DIMCONTINUE	ICL	IMAGECLIP
DDA	DIMDISASSOCIATE	IM	IMAGE

Alias	Command
-IM	-IMAGE
IMP	IMPORT
IN	INTERSECT
INF	INTERFERE
IO	INSERTOBJ
L	LINE
LA	LAYER
-LA	-LAYER
LE	QLEADER
LEN	LENGTHEN
LI	LIST
LINEWEIGHT	LWEIGHT
LO	-LAYOUT
LS	LIST
LT	LINETYPE
-LT	-LINETYPE
LTYPE	LINETYPE
-LTYPE	-LINETYPE
LTS	LTSCALE
LW	LWEIGHT
M	MOVE
MA	MATCHPROP
ME	MEASURE
MI	MIRROR
ML	MLINE
MO	PROPERTIES
MS	MSPACE
MT	MTEXT
MV	MVIEW
O	OFFSET
OP	OPTIONS
ORBIT	3DORBIT
OS	OSNAP
-OS	-OSNAP
P	PAN
-P	-PAN
PA	PASTESPEC
PARTIALOPEN	-PARTIALOPEN
PE	PEDIT
PL	PLINE
PO	POINT
POL	POLYGON
PR	OPTIONS
PRCLOSE	PROPERTIESCLOSE
PROPS	PROPERTIES
PRE	PREVIEW
PRINT	PLOT
PS	PSPACE
PTW	PUBLISHTOWEB
PU	PURGE
-PU	-PURGE
R	REDRAW
RA	REDRAWALL
RE	REGEN
REA	REGENALL
REC	RECTANGLE
REG	REGION
REN	RENAME
-REN	-RENAME
REV	REVOLVE
RM	DDRMODES
RO	ROTATE
RPR	RPREF
RR	RENDER
S	STRETCH
SC	SCALE
SCR	SCRIPT
SE	DSETTINGS
SEC	SECTION
SET	SETVAR
SHA	SHADE
SL	SLICE
SN	SNAP
SO	SOLID
SP	SPELL
SPL	SPLINE
SPE	SPLINEDIT
ST	STYLE
SU	SUBTRACT
T	MTEXT
-T	-MTEXT
TA	TABLET
TH	THICKNESS
TI	TILEMODE
TO	TOOLBAR
TOL	TOLERANCE
TOR	TORUS
TR	TRIM
UC	DDUCS
UCP	DDUCSP
UN	UNITS
-UN	-UNITS
UNI	UNION
V	VIEW
-V	-VIEW
VP	DDVPOINT
-VP	VPOINT
W	WBLOCK
-W	-WBLOCK
WE	WEDGE

Alias	Command	Alias	Command
X	EXPLODE	XL	XLINE
XA	XATTACH	XR	XREF
XB	XBIND	-XR	-XREF
-XB	-XBIND	Z	ZOOM
XC	XCLIP		

APPENDIX 2

AutoCAD Toolbars

This is a listing of all the toolbars in AutoCAD 2002 and the commands each icon executes. The second column lists the basic AutoCAD command while the third includes all options used by the toolbar.

3D Orbit		
3D Pan	3dpan	'_3dpan
3D Zoom	3dzoom	'_3dzoom
3D Orbit	3dorbit	'_3dorbit
3D Continuous Orbit	3dcorbit	'_3dcorbit
3D Swivel	3dswivel	'_3dswivel
3D Adjust Distance	3ddistance	'_3ddistance
3D Adjust Clip Planes	3dclip	'_3dclip
Front Clip On/Off	dview	_dview _all _cl _f
Back Clip On/Off	dview	_dview _all _cl _b
CAD Standards		
Configure Standards	Standards	_Standards
Check Standards	CheckStandards	_CheckStandards
Layer Translate	LayTrans	_LayTrans
Dimension		
Linear Dimension	dimlinear	_dimlinear
Aligned Dimension	dimaligned	_dimaligned
Ordinate Dimension	dimordinate	_dimordinate
Radius Dimension	dimradius	_dimradius
Diameter Dimension	dimdiameter	_dimdiameter
Angular Dimension	dimangular	_dimangular
Quick Dimension	qdim	_qdim
Baseline Dimension	dimbaseline	_dimbaseline
Continue Dimension	dimcontinue	_dimcontinue
Quick Leader	qleader	_qleader
Tolerance	tolerance	_tolerance
Center Mark	dimcenter	_dimcenter
Dimension Edit	dimedit	_dimedit
Dimension Text Edit	dimtedit	_dimtedit
Dimension Update	dimstyle	_-dimstyle _apply
Dimension Style Dropdown		
Dimension Style	dimstyle	'_dimstyle
Draw		
Line	line	_line
Construction Line	xline	_xline
Multiline	mline	_mline

Polyline	pline	_pline
Polygon	polygon	_polygon
Rectangle	rectang	_rectang
Arc	arc	_arc
Circle	circle	_circle
Ellipse	ellipse	_ellipse
Ellipse Arc	ellipse	_ellipse _a
Insert Block	insert	_insert
Make Block	block	_block
Point	point	_point
Hatch	bhatch	_bhatch
Region	region	_region
Multiline Text	mtext	_mtext
Spline	spline	_spline
Inquiry		
Distance	dist	'_dist
Area	area	_area
Region/Mass Properties	massprop	_massprop
List	list	_list
Locate Point	id	'_id
Insert		
Insert Block	insert	_insert
External Reference	xref	_xref
Image	image	_image
Import	import	_import
OLE Object	insertobj	_insertobj
Layouts		
New Layout	layout	_layout _n
Layout from Template	layout	_layout _t
Page Setup	pagesetup	_pagesetup
Display Viewports Dialog	vports	_vports
Modify		
Erase	erase	_erase
Copy Object	copy	_copy
Mirror	mirror	_mirror
Offset	offset	_offset
Array	array	_array
Move	move	_move
Rotate	rotate	_rotate
Scale	scale	_scale
Stretch	stretch	_stretch
Lengthen	lengthen	_lengthen
Trim	trim	_trim
Extend	extend	_extend
Break at Point	break	_break \f \@
Break	break	_break
Chamfer	chamfer	_chamfer
Fillet	fillet	_fillet
Explode	explode	_explode
Modify II		
Draworder	draworder	_draworder
Edit Hatch	hatchedit	_hatchedit
Edit Polyline	pedit	_pedit

Edit Spline	splinedit	_splinedit
Edit Multiline	mledit	_mledit
Edit Attribute	eattedit	_eattedit
Block Attribute Manager	BattMan	_BattMan
Synchronize Attributes	AttSync	_AttSync
Attribute Extract	EAttExt	_EAttExt
Object Properties		
Make Object's Layer Current	molc	_ai_molc
Layers	layer	'_layer
Layer Dropdown		
Layer Previous	LayerP	_LayerP
Color Dropdown		
Linetype Dropdown		
Lineweight Dropdown		
Plot Styles Dropdown		
Object Snap		
Temporary Tracking Point	tt	_tt
Snap to Tangent	tan	_tan
Snap to Perpendicular	per	_per
Snap to Parallel	par	_par
Snap to Insert	ins	_ins
Snap to Node	nod	_nod
Snap to Nearest	nea	_nea
Snap to None	non	_non
Object Snap Settings	dsettings	'_+dsettings 2
Snap From	from	_from
Snap to Endpoint	endp	_endp
Snap to Midpoint	mid	_mid
Snap to Intersection	int	_int
Snap to Apparent Intersect	appint	_appint
Snap to Extension	ext	_ext
Snap to Center	cen	_cen
Snap to Quadrant	qua	_qua
Refedit		
Edit block or Xref	refedit	_refedit;
Add objects to working set		
Remove objects from working set		
Discard changes to reference		
Save back changes to reference		
Reference		
External Reference	xref	_xref
External Reference Attach	xattach	_xattach
External Reference Clip	xclip	_xclip
External Reference Bind	xbind	_xbind
External Reference Clip Frame	xclipframe	_xclipframe 0
Image	image	_image
Image Attach	imageattach	_imageattach
Image Clip	imageclip	_imageclip
Image Adjust	imageadjust	_imageadjust
Image Quality	imagequality	_imagequality
Image Transparency	transparency	_transparency
Image Frame	imageframe	_imageframe

Render		
Hide	hide	_hide
Render	render	_render
Scenes	scene	_scene
Lights	light	_light
Materials	rmat	_rmat
Materials Library	matlib	_matlib
Mapping	setuv	_setuv
Background	background	_background
Fog	fog	_fog
Landscape New	lsnew	_lsnew
Landscape Edit	lsedit	_lsedit
Landscape Library	lslib	_lslib
Render Preferences	rpref	_rpref
Statistics	stats	_stats
Shade		
2D Wireframe	shademode	_shademode _2
3D Wireframe	shademode	_shademode _3
Hidden	shademode	_shademode _h
Flat Shaded	shademode	_shademode _f
Gouraud Shaded	shademode	_shademode _g
Flat Shaded, Edges On	shademode	_shademode _l
Gouraud Shaded, Edges On	shademode	_shademode _o
Solids		
Box	box	_box
Sphere	sphere	_sphere
Cylinder	cylinder	_cylinder
Cone	cone	_cone
Wedge	wedge	_wedge
Torus	torus	_torus
Extrude	extrude	_extrude
Revolve	revolve	_revolve
Slice	slice	_slice
Section	section	_section
Interfere	interfere	_interfere
Setup Drawing	soldraw	_soldraw
Setup View	solview	_solview
Setup Profile	solprof	_solprof
Solids Editing		
Union	union	_union
Subtract	subtract	_subtract
Intersect	intersect	_intersect
Extrude Faces	solidedit	_solidedit _face _extrude
Move Faces	solidedit	_solidedit _face _move
Offset Faces	solidedit	_solidedit _face _offset
Delete Faces	solidedit	_solidedit _face _delete
Rotate Faces	solidedit	_solidedit _face _rotate
Taper Faces	solidedit	_solidedit _face _taper
Copy Faces	solidedit	_solidedit _face _copy
Color Faces	solidedit	_solidedit _face _color
Copy Edges	solidedit	_solidedit _edge _copy
Color Edges	solidedit	_solidedit _edge _color
Imprint	solidedit	_solidedit _body _imprint
Clean	solidedit	_solidedit _body _clean

Separate	solidedit	_solidedit _body _separate
Shell	solidedit	_solidedit _body _shell
Check	solidedit	_solidedit _body _check
Standard Toolbar		
New	new	_new
Open	open	_open
Save	qsave	_qsave
Plot	plot	_plot
Plot Preview	preview	_preview
Find and Replace	find	_find
Cut to Clipboard	cutclip	_cutclip
Copy to Clipboard	copyclip	_copyclip
Paste from Clipboard	pasteclip	_pasteclip
Match Properties	matchprop	'_matchprop
Undo	u	_u
Redo	redo	_redo
Today	Today	_Today
Autodesk Point A	browser	_browser http://pointa.autodesk.com
Meet Now	MeetNow	_MeetNow
Publish to Web	PublishToWeb	_PublishToWeb
eTransmit	etransmit	_etransmit
Insert Hyperlink	hyperlink	_hyperlink
Object Snap		Object Snap Flyout
UCS		UCS flyout
Viewpoint		Viewpoint Flyout
3D Orbit	3dorbit	'_3dorbit
Pan Realtime	pan	'_pan
Zoom Realtime	zoom	'_zoom ;
Zoom	zoom	_zoom
Zoom Previous	zoom	'_zoom _p
AutoCAD DesignCenter	adcenter	'_adcenter
Properties	properties	_properties
Help	help	'_help
Active Assistance	ASSIST	'_ASSIST
Surfaces		
2D Solid	solid	_solid
3D Face	3dface	_3dface
Box	box	_ai_box
Wedge	wedge	_ai_wedge
Pyramid	pyramid	_ai_pyramid
Cone	cone	_ai_cone
Sphere	sphere	_ai_sphere
Dome	dome	_ai_dome
Dish	dish	_ai_dish
Torus	torus	_ai_torus
Edge	edge	_edge
3D Mesh	3dmesh	_3dmesh
Revolved Surface	revsurf	_revsurf
Tabulated Surface	tabsurf	_tabsurf
Ruled Surface	rulesurf	_rulesurf
Edge Surface	edgesurf	_edgesurf

Text

Multiline Text	mtext	_mtext
Single Line Text	dtext	_dtext
Edit Text	ddedit	_ddedit
Find and Replace	find	_find
Text Style	style	'_style
Scale Text	scaletext	_scaletext
Justify Text	justifytext	_justifytext
Convert distance between spaces	spacetrans	'_spacetrans

UCS

UCS	ucs	_ucs
Display UCS Dialog	ucsman	_+ucsman 0
UCS Previous	ucs	_ucs _p
World UCS	ucs	_ucs _w
Object UCS	ucs	_ucs _ob
Face UCS	ucs	_ucs _fa
View UCS	ucs	_ucs _v
Origin UCS	ucs	_ucs _o
Z Axis Vector UCS	ucs	_ucs _zaxis
3 Point UCS	ucs	_ucs _3
X Axis Rotate UCS	ucs	_ucs _x
Y Axis Rotate UCS	ucs	_ucs _y
Z Axis Rotate UCS	ucs	_ucs _z
Apply UCS	ucs	_ucs _apply

UCS II

Display UCS Dialog	ucsman	_+ucsman 0
Move UCS Origin	ucs	_ucs _move
UCS dropdown list		

View

Named Views	view	_view
Top View	view	_-view _top
Bottom View	view	_-view _bottom
Left View	view	_-view _left
Right View	view	_-view _right
Front View	view	_-view _front
Back View	view	_-view _back
SW Isometric View	view	_-view _swiso
SE Isometric View	view	_-view _seiso
NE Isometric View	view	_-view _neiso
NW Isometric View	view	_-view _nwiso
Camera	camera	_camera

Viewports

Display Viewports Dialog	vports	_vports
Single Viewport	vports	_-vports _si
Polygonal Viewport	vports	_-vports _p
Convert Object to Viewport	vports	_-vports _o
Clip Existing Viewport	vpclip	_vpclip

Web

Go Back	hyperlinkBack	'_hyperlinkBack
Go Forward	hyperlinkFwd	'_hyperlinkFwd
Stop Navigation	hyperlinkStop	'_hyperlinkStop
Browse the Web	browser	_browser ;

Zoom		
Zoom Window	zoom	'_zoom _w
Zoom Dynamic	zoom	'_zoom _d
Zoom Scale	zoom	'_zoom _s
Zoom Center	zoom	'_zoom _c
Zoom In	zoom	'_zoom 2x
Zoom Out	zoom	'_zoom .5x
Zoom All	zoom	'_zoom _all
Zoom Extents	zoom	'_zoom _e

APPENDIX 3

List of AutoCAD Commands and Where to Find Them on the Toolbars

The toolbar location is given by toolbar name followed by a number representing where on the toolbar the icon appears, thus an entry of "3D Orbit 7" would be the seventh icon on the 3D orbit toolbar.

Command	Toolbar Location	Tool Tip
3dclip	3D Orbit 7	3D Adjust Clip Planes
3dcorbit	3D Orbit 4	3D Continuous Orbit
3ddistance	3D Orbit 6	3D Adjust Distance
3dface	Surfaces 2	3D Face
3dmesh	Surfaces 12	3D Mesh
3dorbit	3D Orbit 3	3D Orbit
3dorbit	Standard Toolbar 22	3D Orbit
3dpan	3D Orbit 1	3D Pan
3dswivel	3D Orbit 5	3D Swivel
3dzoom	3D Orbit 2	3D Zoom
adcenter	Standard Toolbar 27	AutoCAD DesignCenter
appint	Object Snap 6	Snap to Apparent Intersect
arc	Draw 7	Arc
area	Inquiry 2	Area
array	Modify 5	Array
ASSIST	Standard Toolbar 30	Active Assistance
AttSync	Modify II 8	Synchronize Attributes
background	Render 8	Background
BattMan	Modify II 7	Block Attribute Manager
bhatch	Draw 15	Hatch
block	Draw 13	Make Block
box	Solids 1	Box
box	Surfaces 3	Box
break	Modify 13	Break at Point
break	Modify 14	Break
browser	Standard Toolbar 14	Autodesk Point A
browser	Web 4	Browse the Web
camera	View 12	Camera
cen	Object Snap 8	Snap to Center
chamfer	Modify 15	Chamfer
CheckStandards	CAD Standards 2	Check Standards
circle	Draw 8	Circle
cone	Solids 4	Cone
cone	Surfaces 6	Cone
copy	Modify 2	Copy Object
copyclip	Standard Toolbar 8	Copy to Clipboard

Command	Toolbar Location	Tool Tip
cutclip	Standard Toolbar 7	Cut to Clipboard
cylinder	Solids 3	Cylinder
ddedit	Text 3	Edit Text
dimaligned	Dimension 2	Aligned Dimension
dimangular	Dimension 6	Angular Dimension
dimbaseline	Dimension 8	Baseline Dimension
dimcenter	Dimension 12	Center Mark
dimcontinue	Dimension 9	Continue Dimension
dimdiameter	Dimension 5	Diameter Dimension
dimedit	Dimension 13	Dimension Edit
dimlinear	Dimension 1	Linear Dimension
dimordinate	Dimension 3	Ordinate Dimension
dimradius	Dimension 4	Radius Dimension
dimstyle	Dimension 17	Dimension Style
dimstyle	Dimension 15	Dimension Update
dimtedit	Dimension 14	Dimension Text Edit
dish	Surfaces 9	Dish
dist	Inquiry 1	Distance
dome	Surfaces 8	Dome
draworder	Modify II 1	Draworder
dsettings	Object Snap 17	Object Snap Settings
dtext	Text 2	Single Line Text
dview	3D Orbit 8	Front Clip On/Off
dview	3D Orbit 9	Back Clip On/Off
eattedit	Modify II 6	Edit Attribute
EAttExt	Modify II 9	Attribute Extract
edge	Surfaces 11	Edge
edgesurf	Surfaces 16	Edge Surface
ellipse	Draw 10	Ellipse
ellipse	Draw 11	Ellipse Arc
endp	Object Snap 3	Snap to Endpoint
erase	Modify 1	Erase
etransmit	Standard Toolbar 17	eTransmit
explode	Modify 17	Explode
ext	Object Snap 7	Snap to Extension
extend	Modify 12	Extend
extrude	Solids 7	Extrude
fillet	Modify 16	Fillet
find	Standard Toolbar 6	Find and Replace
find	Text 4	Find and Replace
fog	Render 9	Fog
from	Object Snap 2	Snap From
hatchedit	Modify II 2	Edit Hatch
help	Standard Toolbar 29	Help
hide	Render 1	Hide
hyperlink	Standard Toolbar 18	Insert Hyperlink
hyperlinkBack	Web 1	Go Back
hyperlinkFwd	Web 2	Go Forward
hyperlinkStop	Web 3	Stop Navigation
id	Inquiry 5	Locate Point
image	Insert 3	Image
image	Reference 6	Image
imageadjust	Reference 9	Image Adjust
imageattach	Reference 7	Image Attach
imageclip	Reference 8	Image Clip

Command	Toolbar Location	Tool Tip
imageframe	Reference 12	Image Frame
imagequality	Reference 10	Image Quality
import	Insert 4	Import
ins	Object Snap 13	Snap to Insert
insert	Draw 12	Insert Block
insert	Insert 1	Insert Block
insertobj	Insert 5	OLE Object
int	Object Snap 5	Snap to Intersection
interfere	Solids 11	Interfere
intersect	Solids Editing 3	Intersect
justifytext	Text 7	Justify Text
layer	Object Properties 2	Layers
LayerP	Object Properties 4	Layer Previous
layout	Layouts 1	New Layout
layout	Layouts 2	Layout from Template
LayTrans	CAD Standards 3	Layer Translate
lengthen	Modify 10	Lengthen
light	Render 4	Lights
line	Draw 1	Line
list	Inquiry 4	List
lsedit	Render 11	Landscape Edit
lslib	Render 12	Landscape Library
lsnew	Render 10	Landscape New
massprop	Inquiry 3	Region/Mass Properties
matchprop	Standard Toolbar 10	Match Properties
matlib	Render 6	Materials Library
MeetNow	Standard Toolbar 15	Meet Now
mid	Object Snap 4	Snap to Midpoint
mirror	Modify 3	Mirror
mledit	Modify II 5	Edit Multiline
mline	Draw 3	Multiline
molc	Object Properties 1	Make Object's Layer Current
move	Modify 6	Move
mtext	Draw 17	Multiline Text
mtext	Text 1	Multiline Text
nea	Object Snap 15	Snap to Nearest
new	Standard Toolbar 1	New
nod	Object Snap 14	Snap to Node
non	Object Snap 16	Snap to None
offset	Modify 4	Offset
open	Standard Toolbar 2	Open
pagesetup	Layouts 3	Page Setup
pan	Standard Toolbar 23	Pan Realtime
par	Object Snap 12	Snap to Parallel
pasteclip	Standard Toolbar 9	Paste from Clipboard
pedit	Modify II 3	Edit Polyline
per	Object Snap 11	Snap to Perpendicular
pline	Draw 4	Polyline
plot	Standard Toolbar 4	Plot
point	Draw 14	Point
polygon	Draw 5	Polygon
preview	Standard Toolbar 5	Plot Preview
properties	Standard Toolbar 28	Properties
PublishToWeb	Standard Toolbar 16	Publish to Web
pyramid	Surfaces 5	Pyramid

Command	Toolbar Location	Tool Tip
qdim	Dimension 7	Quick Dimension
qleader	Dimension 10	Quick Leader
qsave	Standard Toolbar 3	Save
qua	Object Snap 9	Snap to Quadrant
rectang	Draw 6	Rectangle
redo	Standard Toolbar 12	Redo
refedit	Refedit 1	Edit block or Xref
region	Draw 16	Region
render	Render 2	Render
revolve	Solids 8	Revolve
revsurf	Surfaces 13	Revolved Surface
rmat	Render 5	Materials
rotate	Modify 7	Rotate
rpref	Render 13	Render Preferences
rulesurf	Surfaces 15	Ruled Surface
scale	Modify 8	Scale
scaletext	Text 6	Scale Text
scene	Render 3	Scenes
section	Solids 10	Section
setuv	Render 7	Mapping
shademode	Shade 1	2D Wireframe
shademode	Shade 2	3D Wireframe
shademode	Shade 3	Hidden
shademode	Shade 4	Flat Shaded
shademode	Shade 5	Gouraud Shaded
shademode	Shade 6	Flat Shaded, Edges On
shademode	Shade 7	Gouraud Shaded, Edges On
slice	Solids 9	Slice
soldraw	Solids 12	Setup Drawing
solid	Surfaces 1	2D Solid
solidedit	Solids Editing 4	Extrude Faces
solidedit	Solids Editing 5	Move Faces
solidedit	Solids Editing 6	Offset Faces
solidedit	Solids Editing 7	Delete Faces
solidedit	Solids Editing 8	Rotate Faces
solidedit	Solids Editing 9	Taper Faces
solidedit	Solids Editing 10	Copy Faces
solidedit	Solids Editing 11	Color Faces
solidedit	Solids Editing 12	Copy Edges
solidedit	Solids Editing 13	Color Edges
solidedit	Solids Editing 14	Imprint
solidedit	Solids Editing 15	Clean
solidedit	Solids Editing 16	Separate
solidedit	Solids Editing 17	Shell
solidedit	Solids Editing 18	Check
solprof	Solids 14	Setup Profile
solview	Solids 13	Setup View
spacetrans	Text 8	Convert distance between spaces
sphere	Solids 2	Sphere
sphere	Surfaces 7	Sphere
spline	Draw 9	Spline
splinedit	Modify II 4	Edit Spline
Standards	CAD Standards 1	Configure Standards
stats	Render 14	Statistics

Command	Toolbar Location	Tool Tip
stretch	Modify 9	Stretch
style	Text 5	Text Style
subtract	Solids Editing 2	Subtract
tabsurf	Surfaces 14	Tabulated Surface
tan	Object Snap 10	Snap to Tangent
Today	Standard Toolbar 13	Today
tolerance	Dimension 11	Tolerance
torus	Solids 6	Torus
torus	Surfaces 10	Torus
transparency	Reference 11	Image Transparency
trim	Modify 11	Trim
tt	Object Snap 1	Temporary Tracking Point
u	Standard Toolbar 11	Undo
ucs	UCS 1	UCS
ucs	UCS 3	UCS Previous
ucs	UCS 4	World UCS
ucs	UCS 5	Object UCS
ucs	UCS 6	Face UCS
ucs	UCS 7	View UCS
ucs	UCS 8	Origin UCS
ucs	UCS 9	Z Axis Vector UCS
ucs	UCS 10	3 Point UCS
ucs	UCS 11	X Axis Rotate UCS
ucs	UCS 12	Y Axis Rotate UCS
ucs	UCS 13	Z Axis Rotate UCS
ucs	UCS 14	Apply UCS
ucs	UCS II 2	Move UCS Origin
ucsman	UCS 2	Display UCS Dialog
ucsman	UCS II 1	Display UCS Dialog
union	Solids Editing 1	Union
view	View 1	Named Views
view	View 2	Top View
view	View 3	Bottom View
view	View 4	Left View
view	View 5	Right View
view	View 6	Front View
view	View 7	Back View
view	View 8	SW Isometric View
view	View 9	SE Isometric View
view	View 10	NE Isometric View
view	View 11	NW Isometric View
vpclip	Viewports 5	Clip Existing Viewport
vports	Layouts 4	Display Viewports Dialog
vports	Viewports 1	Display Viewports Dialog
vports	Viewports 2	Single Viewport
vports	Viewports 3	Polygonal Viewport
vports	Viewports 4	Convert Object to Viewport
wedge	Solids 5	Wedge
wedge	Surfaces 4	Wedge
xattach	Reference 2	External Reference Attach
xbind	Reference 4	External Reference Bind
xclip	Reference 3	External Reference Clip
xclipframe	Reference 5	External Reference Clip Frame
xline	Draw 2	Construction Line
xref	Insert 2	External Reference

Command	Toolbar Location	Tool Tip
xref	Reference 1	External Reference
zoom	Standard Toolbar 24	Zoom Realtime
zoom	Standard Toolbar 25	Zoom
zoom	Standard Toolbar 26	Zoom Previous
zoom	Zoom 1	Zoom Window
zoom	Zoom 2	Zoom Dynamic
zoom	Zoom 3	Zoom Scale
zoom	Zoom 4	Zoom Center
zoom	Zoom 5	Zoom In
zoom	Zoom 6	Zoom Out
zoom	Zoom 7	Zoom All
zoom	Zoom 8	Zoom Extents

Standard AutoCAD Fonts

The following fonts are standard with AutoCAD. To use one of these fonts you must use the STYLE command and assign the given font to a text style. The font which is closest to Single Stroke Gothic is ROMANS.

Complex	a b c d e f g h i j k l m n o p q r s t u v w x y z 1 2 3 4 5 6 7 8 9 0 ` - = \ , . / ; ' [ ]
GothicE	a b c d e f g h i j k l m n o p q r s t u v w x y z 1 2 3 4 5 6 7 8 9 0 ` - = \ , . / ; ' [ ]
GothicG	a b c d e f g h i j k l m n o p q r ſ t u v w x y z 1 2 3 4 5 6 7 8 9 0 ` - = \ , . / ; ' [ ]
GothicI	a b c d e f g h i j k l m n o p q r s t u v w x y z 1 2 3 4 5 6 7 8 9 0 ` - = \ , . / ; ' [ ]
GreekC	α β χ δ ε φ γ η ι ∂ κ λ μ ν ο π ϑ ρ σ τ υ ∈ ω ξ ψ ζ 1 2 3 4 5 6 7 8 9 0 ` - = \ , . / ; ' [ ]
GreekS	α β χ δ ε φ γ η ι ∂ κ λ μ ν ο π ϑ ρ σ τ υ ∈ ω ξ ψ ζ 1 2 3 4 5 6 7 8 9 0 ` - = \ , . / ; ' [ ]
IGES1001	∠ ⏀ ▱ ⌓ ○ // ⌭ ↗ ≡ ⌖ ⌒ ⊥ Ⓜ ⌀ ○ Ⓟ ℄ ◎ Ⓢ ◻ ⊙ △ ◇ ⌖ ⩒ Y 1 2 3 4 5 6 7 8 9 0 ` - = \ , . / ; ' [ ]
IGES1002	⧖ ÷ ≤ ≥ △ √ × ≡ ≠ ∫ ⊃ ∨ ∧ ≈ Σ ↑ ↓ → ← φ θ γ ψ ω λ α 1 2 3 4 5 6 7 8 9 0 ` - = \ , . / ; ' [ ]
IGES1003	∠ ⊥ ▱ ⌓ ○ // ⌭ ↗ ≡ ⌖ ⌒ Ⓛ Ⓜ ⌀ ◻ Ⓟ ℄ ◎ Ⓢ – ⊔ ∨ ∓ ⌖ ⌳ 1 2 3 4 5 6 7 8 9 0 ` - = \ , . / ; ' [ ]
Italic	*a b c d e f g h i j k l m n o p q r s t u v w x y z 1 2 3 4 5 6 7 8 9 0 ` - = \ , . / ; ' [ ]*
Italic8	*a b c d e f g h i j k l m n o p q r s t u v w x y z 1 2 3 4 5 6 7 8 9 0 ` - = \ , . / ; ' [ ]*
ItalicC	*a b c d e f g h i j k l m n o p q r s t u v w x y z 1 2 3 4 5 6 7 8 9 0 ` - = \ , . / ; ' [ ]*
ItalicT	***a b c d e f g h i j k l m n o p q r s t u v w x y z 1 2 3 4 5 6 7 8 9 0 ` - = \ , . / ; ' [ ]***
Monotxt	a b c d e f g h i j k l m n o p q r s t u v w x y z 1 2 3 4 5 6 7 8 9 0 ` - = \ , . / ; ' [ ]
RomanC	a b c d e f g h i j k l m n o p q r s t u v w x y z 1 2 3 4 5 6 7 8 9 0 ` - = \ , . / ; ' [ ]
RomanD	a b c d e f g h i j k l m n o p q r s t u v w x y z 1 2 3 4 5 6 7 8 9 0 ` - = \ , . / ; ' [ ]
RomanS	a b c d e f g h i j k l m n o p q r s t u v w x y z 1 2 3 4 5 6 7 8 9 0 ` - = \ , . / ; ' [ ]
RomanT	**a b c d e f g h i j k l m n o p q r s t u v w x y z 1 2 3 4 5 6 7 8 9 0 ` - = \ , . / ; ' [ ]**
ScriptC	*a b c d e f g h i j k l m n o p q r s t u v w x y z 1 2 3 4 5 6 7 8 9 0 ` - = \ , . / ; ' [ ]*
ScriptS	*a b c d e f g h i j k l m n o p q r s t u v w x y z 1 2 3 4 5 6 7 8 9 0 ` - = \ , . / ; ' [ ]*
Simplex	a b c d e f g h i j k l m n o p q r s t u v w x y z 1 2 3 4 5 6 7 8 9 0 ` - = \ , . / ; ' [ ]
Syastro	✱ ' ' ⊂ ∪ ⊃ ∩ ∈ → ↑ ← ↓ ∂ ∇ ⌒ ' ` ˘ ℵ § † ‡ ∃ £ ® © 1 2 3 4 5 6 7 8 9 0 ` - = \ , . / ; ' [ ]
Symap	↟ ♣ ⌂ ♡ ⁂ · • ∘ ○ ○ ○ ◯ ◯ ◯ ◯ ‖ ⊥ ∠ ∴ ♤ ♡ ◇ ♧ ♧ ♣ 1 2 3 4 5 6 7 8 9 0 ` - = \ , . / ; ' [ ]
Symath	← ↓ ∂ ∇ √ ∫ ∮ ∞ § † ‡ ∃ Π Σ ( ) [ ] { } { } ( ) √ ∫ ≈ ≅ 1 2 3 4 5 6 7 8 9 0 ` – = \ , . / ; ' [ ]
Symeteo	| \ ⟍ ⸝ ı \ ⌒ ⌣ ⌣ ( ) ⌒ ∿ ∿ ∿ ∿ ℓ ∝ ℧ ⊳ φ φ • 1 2 3 4 5 6 7 8 9 0 ` – = \ , . / ; ' [ ]
Symusic	· ⌒ ◡ ○ ○ • ♯ ♮ ♭ – – 𝄽 𝄾 𝄞 𝄢 𝄡 ☉ ☿ ♀ ⊕ ♂ ♃ ♄ ♅ ♆ ♇ 1 2 3 4 5 6 7 8 9 0 ` – = \ , . / ; ' [ ]
Txt	a b c d e f g h i j k l m n o p q r s t u v w x y z 1 2 3 4 5 6 7 8 9 0 ` - = \ , . / ; ' [ ]

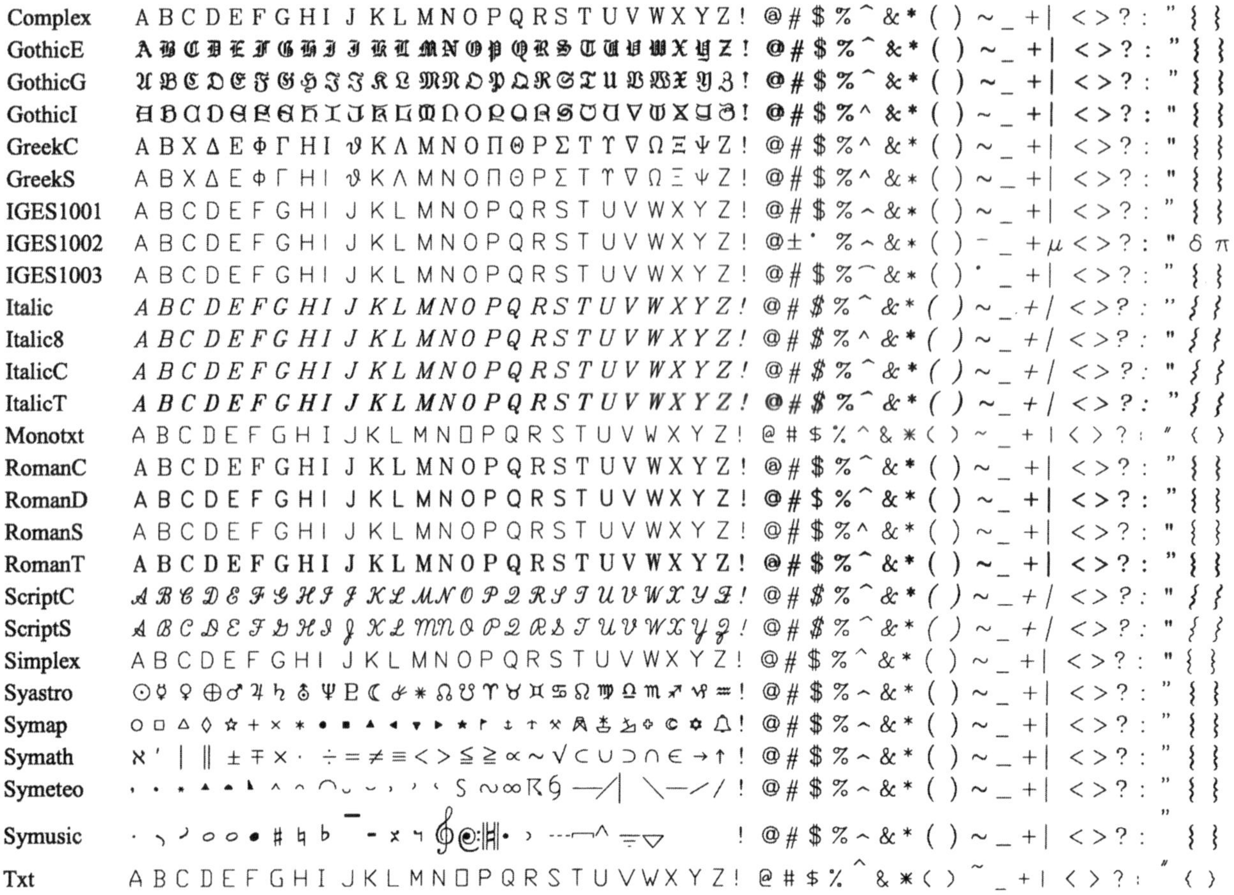

Complex A B C D E F G H I J K L M N O P Q R S T U V W X Y Z ! @ # $ % ^ & * ( ) ~ _ + | < > ? : " { }
GothicE A B C D E F G H I J K L M N O P Q R S T U V W X Y Z ! @ # $ % ^ & * ( ) ~ _ + | < > ? : " { }
GothicG A B C D E F G H I J K L M N O P Q R S T U V W X Y Z ! @ # $ % ^ & * ( ) ~ _ + | < > ? : " { }
GothicI A B C D E F G H I J K L M N O P Q R S T U V W X Y Z ! @ # $ % ^ & * ( ) ~ _ + | < > ? : " { }
GreekC Α Β Χ Δ Ε Φ Γ Η Ι ϑ Κ Λ Μ Ν Ο Π Θ Ρ Σ Τ Υ ∇ Ω Ξ Ψ Ζ ! @ # $ % ^ & * ( ) ~ _ + | < > ? : " { }
GreekS Α Β Χ Δ Ε Φ Γ Η Ι ϑ Κ Λ Μ Ν Ο Π Θ Ρ Σ Τ Υ ∇ Ω Ξ Ψ Ζ ! @ # $ % ^ & * ( ) ~ _ + | < > ? : " { }
IGES1001 A B C D E F G H I J K L M N O P Q R S T U V W X Y Z ! @ # $ % ~ & * ( ) ~ _ + | < > ? : " { }
IGES1002 A B C D E F G H I J K L M N O P Q R S T U V W X Y Z ! @ ± ˙ % ~ & * ( ) ⁻ _ + μ < > ? : " δ π
IGES1003 A B C D E F G H I J K L M N O P Q R S T U V W X Y Z ! @ # $ % ⌒ & * ( ) ˙ _ + | < > ? : " { }
Italic *A B C D E F G H I J K L M N O P Q R S T U V W X Y Z ! @ # $ % ^ & * ( ) ~ _ + / < > ? : " { }*
Italic8 *A B C D E F G H I J K L M N O P Q R S T U V W X Y Z ! @ # $ % ^ & * ( ) ~ _ + / < > ? : " { }*
ItalicC *A B C D E F G H I J K L M N O P Q R S T U V W X Y Z ! @ # $ % ^ & * ( ) ~ _ + / < > ? : " { }*
ItalicT ***A B C D E F G H I J K L M N O P Q R S T U V W X Y Z ! @ # $ % ^ & * ( ) ~ _ + / < > ? : " { }***
Monotxt A B C D E F G H I J K L M N O P Q R S T U V W X Y Z ! @ # $ % ^ & * ( ) ~ _ + | < > ? : " ( )
RomanC A B C D E F G H I J K L M N O P Q R S T U V W X Y Z ! @ # $ % ^ & * ( ) ~ _ + | < > ? : " { }
RomanD A B C D E F G H I J K L M N O P Q R S T U V W X Y Z ! @ # $ % ^ & * ( ) ~ _ + | < > ? : " { }
RomanS A B C D E F G H I J K L M N O P Q R S T U V W X Y Z ! @ # $ % ^ & * ( ) ~ _ + | < > ? : " { }
RomanT **A B C D E F G H I J K L M N O P Q R S T U V W X Y Z ! @ # $ % ^ & * ( ) ~ _ + | < > ? : " { }**
ScriptC *A B C D E F G H I J K L M N O P Q R S T U V W X Y Z ! @ # $ % ^ & * ( ) ~ _ + / < > ? : " { }*
ScriptS *A B C D E F G H I J K L M N O P Q R S T U V W X Y Z ! @ # $ % ^ & * ( ) ~ _ + / < > ? : " { }*
Simplex A B C D E F G H I J K L M N O P Q R S T U V W X Y Z ! @ # $ % ^ & * ( ) ~ _ + | < > ? : " { }
Syastro ! @ # $ % ~ & * ( ) ~ _ + | < > ? : " { }
Symap ! @ # $ % ~ & * ( ) ~ _ + | < > ? : " { }
Symath ! @ # $ % ~ & * ( ) ~ _ + | < > ? : " { }
Symeteo ! @ # $ % ~ & * ( ) ~ _ + | < > ? : " { }
Symusic ! @ # $ % ~ & * ( ) ~ _ + | < > ? : " { }
Txt A B C D E F G H I J K L M N O P Q R S T U V W X Y Z ! @ # $ % ^ & * ( ) ~ _ + | < > ? : " ( )

APPENDIX 5

Configuration Issues with AutoCAD

There are a few basic parameters which AutoCAD uses to determine the format and style of some of the displays you will see as you are working with AutoCAD. This appendix will explain some of the parameters which may need to be changed to get results as shown in the text. This is not an exhaustive list of all parameters, but will discuss the more common ones. To change a parameter, simply type the parameter name at the command line and AutoCAD will request a new value.

Parameter	Suggested Value	AutoCAD Default	Description
STARTUPTODAY	0	1	Controls the initial startup dialog box. A value of 0 is less confusing for a beginning student and is the format used in compiling this text.
CURSORSIZE	100	5	Controls the size of the crosshairs. The value is the percent of the screen which the crosshairs cover.
UCSFOLLOW	0	0	Controls whether a new viewpoint is created to yield a plan view every time a new coordinate system is created.
WORLDVIEW	1	1	All viewpoints are relative to the World Coordinate System.
DIMASSOC	2	2	Controls the format used when dimensioning.
EXPERT	0	0	Controls the level of messages shown. A setting of zero shows all messages.
FILEDIA	1	1	Controls the display of dialog boxes for file input/output.
CMDDIA	1	1	Will suppress some dialog boxes in favor of command line input.
PICKFIRST	1	1	Will allow pre-selection of objects using grips.
UCSICON	0 - for 2D 3 - for 3D	3	Controls whether the UCSICON will be displayed and where. Zero is not displayed, three is displayed at the origin.

INDEX

D

E

F

G

H

I

K

L

M

N

O

P